ADDISON WESLEY

Math
Makes Sense

3

Practice and Homework Book

Teacher's Edition

Authors

Peggy Morrow   Maggie Martin Connell

PEARSON

Addison
Wesley

Publishing Team
Enid Haley
Lesley Haynes
Tricia Carmichael
Marg Bukta
Lynne Gulliver
Susan Lishman
Stephanie Cox
Kaari Turk
Judy Wilson

Publisher
Claire Burnett

Elementary Math Team Leader
Anne-Marie Scullion

Product Manager
Nashaant Sanghavi

Design
Word & Image Design Studio Inc.

Typesetting
Computer Composition of Canada Inc.

ISBN 0-321-21842-6

Printed and bound in Canada.

1 2 3 4 5 -- WC -- 08 07 06 05 04

PEARSON

Addison Wesley

Contents

To the Teacher

This Practice and Homework Book provides reinforcement of the concepts and skills explored in the *Addison Wesley Math Makes Sense 3* program.

There are two sections in the book. The first section follows the sequence of *Math Makes Sense 3* Student Book. It is intended for use throughout the year as you teach the program. A two-page spread supports the content of each numbered lesson in the Student Book, other than Strategies Toolkit lessons.

In each Lesson:

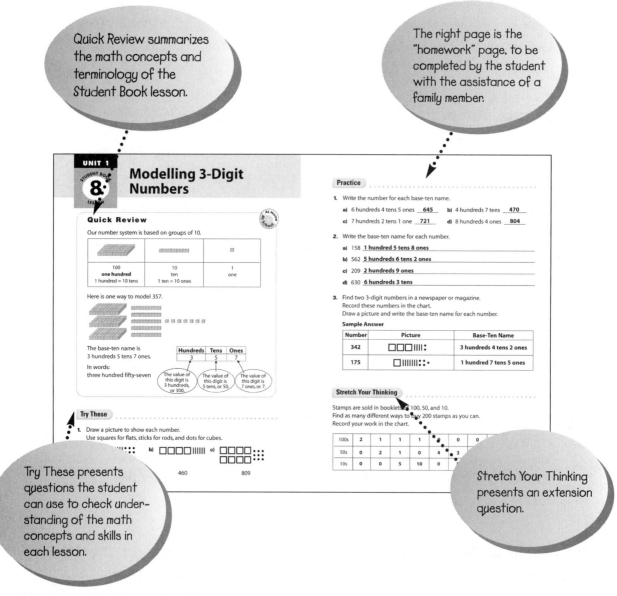

Quick Review summarizes the math concepts and terminology of the Student Book lesson.

The right page is the "homework" page, to be completed by the student with the assistance of a family member.

Try These presents questions the student can use to check understanding of the math concepts and skills in each lesson.

Stretch Your Thinking presents an extension question.

The second section of the book, on perforated pages 197 to 208, consists of 3 four-page Math At Home sections. These fun pages contain intriguing activities, puzzles, rhymes, and games in a magazine format to encourage home involvement. The content makes each section particularly suitable to remove, fold, and send home as an eight-page magazine after the student has completed Units 3, 7, and 11, respectively.

To the Family

The homework pages of this book will help your child practise the math concepts and skills that have been explored in the classroom. As you assist your child to complete each page, you have an opportunity to talk about the math and to become involved in your child's learning.

The left page of a two-page spread always contains a summary of the main concepts and terminology of the lesson that you and your child can use to review the work done in class. The right page contains practice closely linked to the content of the left page.

Here are some ways you can help:
- With your child, read over the Quick Review. Encourage your child to talk about the content and explain it to you in his or her own words.
- Read the instructions with (or for) your child to ensure your child understands what to do.
- Encourage your child to explain his or her thinking as each page is completed.
- Some of the pages require specific materials. You may wish to gather items such as a centimetre ruler, index cards, a measuring tape, scissors, two dice or cubes numbered from 1 to 6, and paper clips.

These homework pages are intended to be enjoyable—many of the Practice sections contain games that will also improve your child's math skills. You may have other ideas for activities your child can share with the rest of the class.

This math workbook will be sent home frequently throughout the year. Please help your child complete the assigned work. Make sure the book is returned promptly.

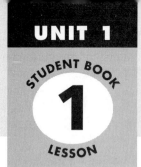

Patterns in a Hundred Chart

Quick Review

Here are some of the patterns in a hundred chart.

➤ Across any row, the numbers increase by 1.

➤ Down any column, the numbers increase by 10. The ones digits are the same and the tens digits increase by 1.

➤ In any diagonal that goes down to the left, the digits add to the same number. The tens digit increases by 1 and the ones digit decreases by 1.

	diagonal		column	

row

1	2	3	4	5	6	7	8	9	10
11	12	13	14	15	16	17	18	19	20
21	22	23	24	25	26	27	28	29	30
31	32	33	34	35	36	37	38	39	40
41	42	43	44	45	46	47	48	49	50
51	52	53	54	55	56	57	58	59	60
61	62	63	64	65	66	67	68	69	70
71	72	73	74	75	76	77	78	79	80
81	82	83	84	85	86	87	88	89	90
91	92	93	94	95	96	97	98	99	100

Try These

Use the hundred chart.

Sample Answers

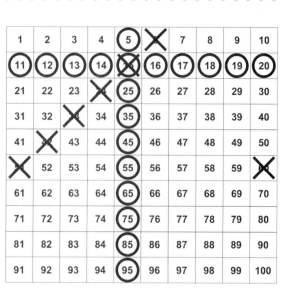

1. a) Circle all the numbers in one row. What pattern do you see?

   The numbers increase by 1.

   b) Circle all the numbers in one column What pattern do you see?

   The numbers increase by 10.

   The ones digits are the same.

   The tens digits increase by 1.

   c) Put an X on all the numbers where the digits add to 6. Where are these numbers?

   Six to 51 are on a diagonal, and 60 is on its own.

Use the hundred chart. Use a different colour for each question.

1. Colour the diagonal that starts at 31 and goes down to the right. What pattern do you see?

   The tens and ones digits both

   increase by 1.

2. Colour the diagonal that starts at 6 and goes down to the left. What pattern do you see?

   The numbers increase by 9.

Sample Answer

1	2	3	4	5	6	7	8	9	10
11	12	13	14	15	16	17	18	19	20
21	22	23	24	25	26	27	28	29	30
31	32	33	34	35	36	37	38	39	40
41	42	43	44	45	46	47	48	49	50
51	52	53	54	55	56	57	58	59	60
61	62	63	64	65	66	67	68	69	70
71	72	73	74	75	76	77	78	79	80
81	82	83	84	85	86	87	88	89	90
91	92	93	94	95	96	97	98	99	100

3. Choose any diagonal that goes down to the left.
   Sample Answers

   a) Write the numbers in the diagonal. **8, 17, 26, 35, 44, 53, 62, 71**

   b) What pattern do you see when you add the digits in each number?

   The digits add to 8.

Stretch Your Thinking

Describe the patterns you see in the rows, columns, and diagonals that go to the left.

1	2	3	4	5	6	7	8	9	10	11	12	13	14	15	16	17	18	19	20
21	22	23	24	25	26	27	28	29	30	31	32	33	34	35	36	37	38	39	40
41	42	43	44	45	46	47	48	49	50	51	52	53	54	55	56	57	58	59	60
61	62	63	64	65	66	67	68	69	70	71	72	73	74	75	76	77	78	79	80
81	82	83	84	85	86	87	88	89	90	91	92	93	94	95	96	97	98	99	100

Across rows, the numbers increase by 1.

Down columns, the numbers increase by 20.

Down the diagonals that go left, the sums of the digits increase by 1.

Counting on a Hundred Chart

Quick Review

You can use patterns on a hundred chart to count.

➤ Start at 36. Count on by 5s.

31	32	33	34	35	36	37	38	39	40
41	42	43	44	45	46	47	48	49	50
51	52	53	54	55	56	57	58	59	60

➤ Start at 75. Count back by 5s.

51	52	53	54	55	56	57	58	59	60
61	62	63	64	65	66	67	68	69	70
71	72	73	74	75	76	77	78	79	80

Note the pattern in the ones digits: 6, 1, 6, 1, 6, …

Note the pattern in the ones digits: 5, 0, 5, 0, 5, …

➤ Start at 17. Count on by 10s.

The ones digit repeats.
The tens digit increases by 1.

11	12	13	14	15	16	17	18	19	20
21	22	23	24	25	26	27	28	29	30
31	32	33	34	35	36	37	38	39	40
41	42	43	44	45	46	47	48	49	50
51	52	53	54	55	56	57	58	59	60
61	62	63	64	65	66	67	68	69	70

Try These

Use a different colour for each pattern.

1. Start at 47.
   Count on by 5s.

2. Start at 51.
   Count on by 10s.

3. Start at 96.
   Count back by 10s to 46.

4. Start at 39.
   Count back by 2s to 21.

1	2	3	4	5	6	7	8	9	10
11	12	13	14	15	16	17	18	19	20
21	22	23	24	25	26	27	28	29	30
31	32	33	34	35	36	37	38	39	40
41	42	43	44	45	46	47	48	49	50
51	52	53	54	55	56	57	58	59	60
61	62	63	64	65	66	67	68	69	70
71	72	73	74	75	76	77	78	79	80
81	82	83	84	85	86	87	88	89	90
91	92	93	94	95	96	97	98	99	100

Practice

Colour on the hundred chart.
Start at any number for each count.
Sample Answers

1. Use blue. Count on by 5s.
   Describe the pattern.

   The ones digit alternates

   between two numbers.

2. Use red. Count on by 10s.
   Describe the pattern.

   The ones digit repeats.

   The tens digit increases by 1.

1	2	3	4	5	6	7	8	9	10
11	12	13	14	15	16	17	18	19	20
21	22	23	24	25	26	27	28	29	30
31	32	33	34	35	36	37	38	39	40
41	42	43	44	45	46	47	48	49	50
51	52	53	54	55	56	57	58	59	60
61	62	63	64	65	66	67	68	69	70
71	72	73	74	75	76	77	78	79	80
81	82	83	84	85	86	87	88	89	90
91	92	93	94	95	96	97	98	99	100

☐ = green   ☐ = blue   ☐ = red

3. Use green. Count back by 2s.
   What do you notice about the numbers?

   They are all even or all odd,

   depending on where you started.

Stretch Your Thinking

1. Suppose you start at 63 and count on to 93.
   a) If you count by 2s, would you say 70? 75? Why or why not?

   I would not say 70, but I would say 75. The pattern in the ones digits

   would be 3, 5, 7, 9, 1, 3, 5, 7, 9, 1, ...

   b) If you count by 5s, would you say 75? 83? Why or why not?

   I would not say 75, but I would say 83. The pattern in the ones

   digits would be 3, 8, 3, 8, 3, 8, ...

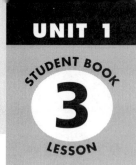

Counting on a Number Line

At Home
At School

Quick Review

You can use a number line to count.

➤ Start at 4. Count on by 5s.

Note the pattern in the ones digits: 4, 9, 4, 9, …

➤ Start at 73. Count back by 10s.

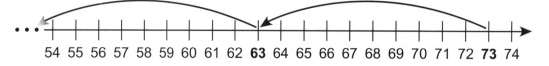

Note the pattern in the ones digits: 3, 3, 3, 3, 3, …

Try These

1. Start at 40. Count on by 2s.

2. Start at 65. Count back by 5s.

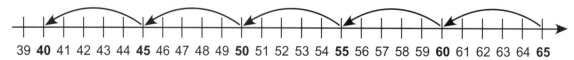

3. What pattern do you see in question 2?

   The pattern in the ones digits is 5, 0, 5, 0, 5, 0, …

1. Start at 26. Count on by 5s. What is the pattern?

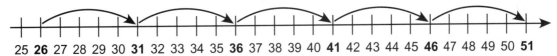

25 **26** 27 28 29 30 **31** 32 33 34 35 **36** 37 38 39 40 **41** 42 43 44 45 **46** 47 48 49 50 **51**

The pattern in the ones digits is 6, 1, 6, 1, 6, 1, ...

2. **a)** Start at 39. Count back by 2s. What is the pattern?

14 **15** 16 **17** 18 **19** 20 **21** 22 **23** 24 **25** 26 **27** 28 **29** 30 **31** 32 **33** 34 **35** 36 **37** 38 **39** 40

The pattern in the ones digits is 9, 7, 5, 3, 1, 9, 7, 5, 3, 1, ...

b) How do you know which numbers come next in the pattern?

Subtract 2 or find the next smaller odd number.

c) Which number in the pattern comes before 39? How do you know?

41; it is the next odd number greater than 39.

3. Fill in the missing numbers.
What are you counting by each time?

a) 53, 48, 43, **38** , **33**

Count back by 5s.

b) 63, 56, 49, **42** , **35**

Count back by 7s.

c) 25, 28, 31, **34** , **37**

Count on by 3s.

d) 39, 43, 47, **51** , **55**

Count on by 4s.

Stretch Your Thinking

Suppose you had a 1 to 100 number line.
How many jumps would you make on it to show counting on by 5s
starting at 11? Explain.

I would make 17 jumps. The pattern in the ones digits is 1, 6, 1, 6, 1, 6, ...

There are 18 numbers starting at 11 and going to 100 in this pattern.

Comparing Numbers on a Number Line

At Home At School

Quick Review

You can estimate where to place numbers on a number line.

To place 82 on this number line:

Think: 82 is between 80 and 85.

82 is closer to 80 than to 85.
Estimate.
Mark a dot between 80 and 85, closer to 80.

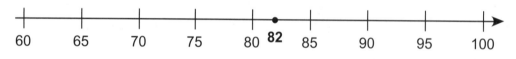

Try These

1. Draw a dot on the number line for each number.
   Label the dot with the number.
   a) 33, 39, 46, 54

   b) 26, 34, 38, 42

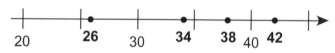

2. Estimate the number for each letter.

Sample Answers

A ___15___   B ___21___   C ___37___

1. Draw a dot on the number line for each number.
    Label the dot with the number.

    a) 54, 69, 76, 81

    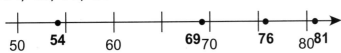

    b) 12, 14, 17, 19

    c) 50, 75, 90, 98

    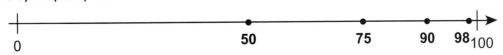

2. Work with a partner.
    Each partner draws 2 dots on the first number line.
    Then each partner labels his or her partner's dots with the numbers.
    Repeat with the other number lines.
    Sample Answers

    a)

    b)

    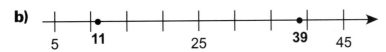

    c)

Stretch Your Thinking

Use this number line to find the secret number.

The number is closer to 60 than to 70. It is one number closer to 65 than 68 is.

What is the secret number? **63**

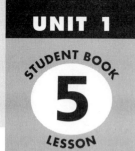

UNIT 1
Grouping and Counting to 100

Quick Review

To count objects,
you can group them into sets.

Here is one way to count a
collection of Snap Cubes.
Make towers of 10 Snap Cubes.

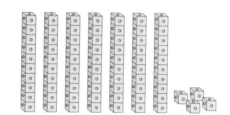

There are 7 towers of 10 Snap Cubes
and 4 leftover cubes.
7 tens and 4 more make 74.
So, there are 74 Snap Cubes.

Grouping objects in tens makes it easier to count them.

Try These

1.

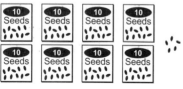

_____**4**_____ tens _____**7**_____ ones

_____**47**_____ cubes

2.

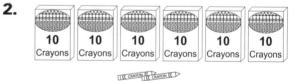

_____**6**_____ tens _____**2**_____ ones

_____**62**_____ crayons

3.

_____**8**_____ tens _____**5**_____ ones

_____**85**_____ seeds

4.

_____**5**_____ tens _____**1**_____ ones

_____**51**_____ balloons

1. Estimate how many stars. My estimate: __**Sample Answer: about 70**__
   Colour the stars. Make each group of ten a different colour. Then count.

   Answer: 7 different sets of coloured stars and 5 stars not coloured

   My count: __**75 stars**__

2. How many craft sticks?

   a)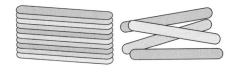

   ___14___

   b)

   ___7___

   c)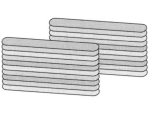

   ___20___

   d)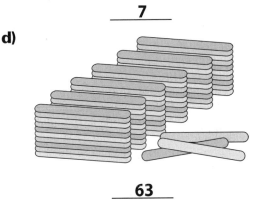

   ___63___

Stretch Your Thinking

1. Suppose all the craft sticks in question 2 were put together.

   a) Add the tens together. How many do you get? __**9 tens**__

   b) Now add the ones. How many do you get? __**14 ones**__

   c) What number is this? _____**104**_____

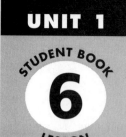
Modelling 2-Digit Numbers

Quick Review

At Home
At School

Here are some different ways to model 26 with Base Ten Blocks.

26 ones

1 ten 16 ones

2 tens 6 ones

When the model has no more
than 9 ones, you can write
the number on a place-value chart.

Tens	Ones
2	6

The value of
this digit is
2 tens, or 20.

The value of
this digit is
6 ones, or 6.

Place value helps us
to understand numbers.
The *value* of each digit depends on
its *place* in the number.

The **base-ten name** for 26 is 2 tens 6 ones. In words: twenty-six

Try These

1. Write each number.

   a)

   b)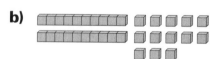

   _____45_____   _____33_____

2. Write the value of each underlined digit.

   a) 2<u>7</u>   __7__   b) <u>8</u>5   __80__   c) <u>1</u>8   __10__   d) 2<u>1</u>   __1__

1. Complete the chart. Use squares for flats, sticks for rods, and dots for cubes.

Number	Picture	Base-Ten Name	Words
45	‖‖‖ :∶•	4 tens 5 ones	**forty-five**
62	▭▭▭▭▭ ▯ ▯	6 tens 2 ones	**sixty-two**
28	‖ ∶∶∶∶	2 tens 8 ones	twenty-eight

2. Draw pictures of Base Ten Blocks. Show each number 3 different ways.
   Sample Answers

25	‖ ∶∶•	‖ ∶∶∶∶∶	⬤⬤⬤⬤⬤ ⬤⬤⬤⬤⬤ ⬤⬤⬤⬤⬤ ⬤⬤⬤⬤⬤ ⬤⬤⬤⬤⬤
41	‖‖‖‖•	‖‖‖ ∶∶∶∶∶•	‖ ∶∶∶∶∶∶∶∶∶∶

1. How many numbers between 1 and 100 have 5 ones? Explain.

   <u>**Ten; there is one for each possible tens digit, from 0 to 9.**</u>

2. How many numbers between 1 and 100 have 5 tens? Explain.

   <u>**Ten; there is one for each possible ones digit, from 0 to 9.**</u>

Ordinal Numbers

Quick Review

Numbers like 56th, 42nd, 71st, and 63rd are **ordinal numbers**.
They tell us the position of an item in a list.

To find the ordinal number of an item:
• Start at 1.
• Count until you reach the item.
• Name the ordinal that goes with the last number of your count.

Find the ordinal number of the striped ball.
Start at the left. Count: 1, 2, 3, 4, 5, 6, 7, 8, 9, 10, 11, 12

The striped ball is 12th.

Try These

1. **a)** Circle the 6th apple.
    b) Draw a worm in the 17th apple.
    c) Colour the 10th apple.
    d) Draw an X on the 18th apple.

2. Use the hundred chart.
    a) Circle the 49th even number.
    b) Place an X on the 6th number
        with a 2 in the ones place.
    c) Start at 1 and count by 2s.
        Colour the 35th number you say.

1	2	3	4	5	6	7	8	9	10
11	12	13	14	15	16	17	18	19	20
21	22	23	24	25	26	27	28	29	30
31	32	33	34	35	36	37	38	39	40
41	42	43	44	45	46	47	48	49	50
51	52	53	54	55	56	57	58	59	60
61	62	63	64	65	66	67	68	69	70
71	72	73	74	75	76	77	78	79	80
81	82	83	84	85	86	87	88	89	90
91	92	93	94	95	96	97	98	99	100

Use the chart below.
1. Draw a star in the 47th square
2. Draw an apple in the 33rd square.
3. Colour the 17th square black.
4. Print your initials in the 20th even-numbered square.

5. Which comes first, the apple or your initials? **the apple**
6. Start counting at the square with the apple in it. (Make this your first square.)

   Now what number square are your initials in? **8th**
7. Draw a happy face in any empty square.

   What is the ordinal number of the square? **Sample Answer: 13th**

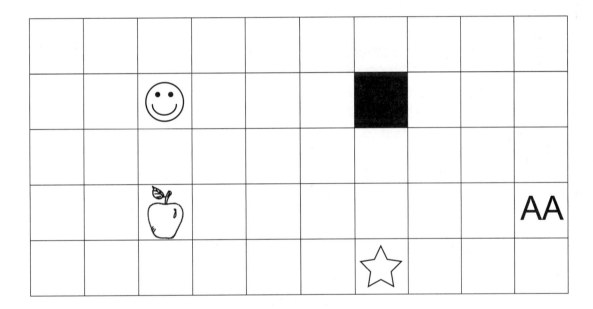

Stretch Your Thinking

1. June 21st is the first day of summer.

   a) Which date will the 40th day of summer be? **July 30th**

   b) Which date will the 60th day of summer be? **August 19th**

   c) August 31st is the _____**72nd**_____ day of summer.

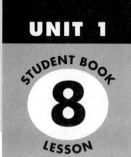

Modelling 3-Digit Numbers

Quick Review

Our number system is based on groups of 10.

100 **one hundred** 1 hundred = 10 tens	10 ten 1 ten = 10 ones	1 one

Here is one way to model 357.

The base-ten name is
3 hundreds 5 tens 7 ones.

In words:
three hundred fifty-seven

Hundreds	Tens	Ones
3	5	7

The value of this digit is 3 hundreds, or 300.

The value of this digit is 5 tens, or 50.

The value of this digit is 7 ones, or 7.

Try These

1. Draw a picture to show each number.
   Use squares for flats, sticks for rods, and dots for cubes.

   a) □□ ||||| ⠿   256

   b) □□□□ ||||| 460

   c) □□□ / □□□ ⠿⠿   809

1. Write the number for each base-ten name.

   a) 6 hundreds 4 tens 5 ones ___**645**___   **b)** 4 hundreds 7 tens ___**470**___

   c) 7 hundreds 2 tens 1 one ___**721**___   **d)** 8 hundreds 4 ones ___**804**___

2. Write the base-ten name for each number.

   a) 158 **1 hundred 5 tens 8 ones**

   b) 562 **5 hundreds 6 tens 2 ones**

   c) 209 **2 hundreds 9 ones**

   d) 630 **6 hundreds 3 tens**

3. Find two 3-digit numbers in a newspaper or magazine.
   Record these numbers in the chart.
   Draw a picture and write the base-ten name for each number.

   Sample Answer

Number	Picture	Base-Ten Name
342	□□□IIII:	3 hundreds 4 tens 2 ones
175	□IIIIIII:∵•	1 hundred 7 tens 5 ones

Stamps are sold in booklets of 100, 50, and 10.
Find as many different ways to buy 200 stamps as you can.
Record your work in the chart.

100s	2	1	1	1	0	0	0	0	0
50s	0	2	1	0	4	3	2	1	0
10s	0	0	5	10	0	5	10	15	20

UNIT 1

LESSON 9

Extending Hundred Chart Patterns

Quick Review

You can use patterns on any hundred chart to count.

➤ Start at 604. Count on by 10s.

601	602	603	**604**	605	606	607	608	609	610
611	612	613	**614**	615	616	617	618	619	620
621	622	623	**624**	625	626	627	628	629	630
631	632	633	**634**	635	636	637	638	639	640
641	642	643	**644**	645	646	647	648	649	650

The ones digit repeats.
The tens digit increases by 1.

➤ Start at 811. Count on by 5s.

801	802	803	804	805	806	807	808	809	810
811	812	813	814	815	**816**	817	818	819	820
821	822	823	824	825	**826**	827	828	829	830
831	832	833	834	835	**836**	837	838	839	840
841	842	843	844	845	**846**	847	848	849	850

Note the pattern in the ones digits: 1, 6, 1, 6, 1, 6, …
Note the pattern in the tens digits: 1, 1, 2, 2, 3, 3, …

Try These

Use the chart to help you.

1. Start at 744.
   Count on by 5s to 769.

   <u>**744, 749, 754, 759, 764, 769**</u>

2. Start at 732.
   Count on by 10s to 792.

   <u>**732, 742, 752, 762, 772, 782, 792**</u>

701	702	703	704	705	706	707	708	709	710
711	712	713	714	715	716	717	718	719	720
721	722	723	724	725	726	727	728	729	730
731	732	733	734	735	736	737	738	739	740
741	742	743	744	745	746	747	748	749	750
751	752	753	754	755	756	757	758	759	760
761	762	763	764	765	766	767	768	769	770
771	772	773	774	775	776	777	778	779	780
781	782	783	784	785	786	787	788	789	790
791	792	793	794	795	796	797	798	799	800

1. Start at 625.

  a) Count on 10 times by 25s.

  625, 650, 675, 700, 725, 750, 775, 800, 825, 850, 875

  b) Count on 10 times by 2s.

  625, 627, 629, 631, 633, 635, 637, 639, 641, 643, 645

2. Count back by 100s.

  a) Start at 999.

  999, 899, 799, 699, 599, 499, 399, 299, 199, 99

  b) Start at 713.

  713, 613, 513, 413, 313, 213, 113, 13

  c) How did you know when to stop?

  Sample Answer: When the number remaining was less than 100

3. Use counting to solve the problem.
Jan has 375 stickers.
Each page in her sticker book holds 25 stickers.

How many pages will Jan's stickers take? **15 pages**
Use words, numbers, or pictures to explain.
**Sample Answer: Count by 25s: 25, 50, 75, 100, 125, 150, 175, 200,
225, 250, 275, 300, 325, 350, 375**

Stretch Your Thinking

Suppose you started at 875 and counted on to 1000.
What might you be counting by?
Give as many answers as you can.

125, 25, 5, or 1

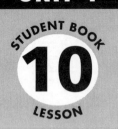
Comparing and Ordering Numbers

Quick Review

You can use place value to **compare** and **order** numbers.

➤ To compare 524 and 528:

1. Compare hundreds.

524
528

Both have 5 hundreds.

2. Compare tens.

5**2**4
5**2**8

Both have 2 tens.

3. Compare ones.

52**4**
52**8**

4 ones are less than 8 ones.

So, 524 **is less than** 528   and   528 **is greater than** 524.

524 < 528   528 > 524

➤ To order 846, 597, and 848, compare each digit.

Hundreds	Tens	Ones
8	4	6
5	9	7
8	4	8

597 has the fewest hundreds, so it is the least number.
848 and 846 have the same number of hundreds and tens.
846 has fewer ones.
So, 846 < 848

The order from least to greatest is 597, 846, 848.
The order from greatest to least is 848, 846, 597.

Try These

1. Write < or > to make a true statement.

a) 845 < 863   **b)** 714 > 703   **c)** 452 > 396

2. Order the numbers from greatest to least.

a) 584, 435, 581   **584, 581, 435**

b) 870, 973, 970   **973, 970, 870**

Practice

1. Circle the greatest number.
   a) 573
      68
      329
      (592)
   b) 925
      (936)
      919
      931
   c) 608
      680
      (724)
      691
   d) 357
      624
      (639)
      620

2. Order the numbers from least to greatest.

   a) 826, 527, 504, 817 **504, 527, 817, 826**

   b) 634, 700, 629, 701 **629, 634, 700, 701**

   c) 358, 324, 196, 238 **196, 238, 324, 358**

3. Write a number to make each statement true.
   Sample Answers

   a) 445 > **425**   b) 799 < **800**   c) 704 < **850**   d) 628 < **641**

4. Use the digits 4, 9, and 6. Make as many 3-digit numbers as you can. Order the numbers from least to greatest.

   469, 496, 649, 694, 946, 964

5. The chart shows how far some students travelled on their holidays. Who travelled:

Name	Distance Travelled
David	825 km
Serena	850 km
Mabel	990 km
Enrique	900 km

   a) the greatest distance? **David**

   b) the least distance? **Mabel**

   c) further than David but not as far as Enrique? **Serena**

Stretch Your Thinking

Use the clues to find the mystery number.
➤ The number is less than 800 but greater than 780.
➤ It has 8 more tens than ones. **791**

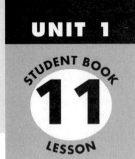

Showing Numbers in Many Ways

Quick Review

Here are different ways to show 340.

Picture: ⬜⬜⬜ ||||

Base-ten name: 3 hundreds 4 tens

Base Ten Blocks

Place-value chart:

Hundreds	Tens	Ones
3	4	0

Standard form: 340

You can use Base Ten Blocks to show 340 in different ways:

Try These

1. Write each number in standard form.

   a) 7 hundreds 4 tens 6 ones   **746**

   b) 8 tens   **80**

   c) 9 hundreds 8 tens 3 ones   **983**

   d) 5 hundreds 2 ones   **502**

2. Write the base-ten name for each number.

   a) 627 **6 hundreds 2 tens 7 ones**

   b) 209 **2 hundreds 9 ones**

   c) 463 **4 hundreds 6 tens 3 ones**

1. Draw a picture to show each number. Use the fewest Base Ten Blocks.

a)	b)	c)
521	309	264

2. Draw a picture of Base Ten Blocks to show 421 in 3 different ways.
   Sample Answer

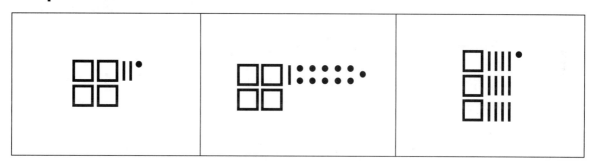

3. Draw a new picture for each number using the fewest blocks.
   Then write each number in standard form.

Picture	Picture	Standard Form
		643
		472

Draw a picture of Base Ten Blocks.
Show 315 using exactly 36 blocks.

Sample Answer

How Much Is 1000?

At Home
At School

Quick Review

Ms. Henry has 10 bags of counters.

Each bag has 100 counters.

To find how many counters Ms. Henry has, you can count by 100s:

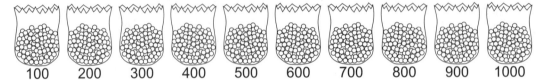

100 200 300 400 500 600 700 800 900 1000

Ten hundreds make 1 **thousand**.

Sammy wants to find out how many yogurt containers 1000 counters
will fill. Here is how he does it:

First Sammy
estimates.

Then he fills a
container with
counters.

Sammy counts the
counters.
There are 104.

Sammy thinks: It took 104 counters to fill one container.

104 is more than 100.

10 hundreds make 1000.

So, 1000 counters will fill fewer than 10 containers.

Try These

1. Draw pictures of Base Ten Blocks to show 1000 in 2 different ways.
   Sample Answer

1. Are there more than 1000 or fewer than 1000:

    a) hairs on a horse? **more** _____

    b) grains of sand on the beach? **more** _____

    c) left-handed students in your school? **fewer** _____

    d) fingers and toes in your classroom? **fewer** _____

Sample Answers

2. Name 3 places where you might see 1000 people.

    At a baseball game, at the mall, at an airport _____

3. Ms. Mansfield is making geoboards for the students in her class. Each geoboard takes 100 pins. How many geoboards can Ms. Mansfield make with 1000 pins? Show how you know.

    Count by 100s: ☐1 ☐2 ☐3 ☐4 ☐5 ☐6 ☐7 ☐8 ☐9 ☐10

                      100 200 300 400 500 600 700 800 900 1000

    So, Ms. Mansfield can make 10 geoboards.

4. Pumpkin seeds come in packages of 50. Mr. Conrad bought 1000 seeds. How many packages did he buy? Use pictures, numbers, or words to explain.

    He bought 20 packages. When you count by 50s, 1000 is the 20th number.

Stretch Your Thinking

Find as many ways as you can to buy exactly 1000 paper clips.

500	2	1	1	0	0	0
250	0	2	0	4	2	0
100	0	0	5	0	5	10

Rounding Numbers

Quick Review

You can use a number line to **round** numbers.

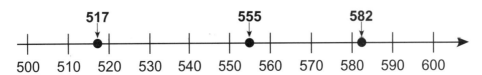

➤ To round to the *nearest ten*

517 is between 510 and 520, but closer to 520.
So, 517 rounds to 520.

555 is halfway between 550 and 560.
When a number is halfway between 2 tens, round to the greater 10.
So, 555 rounds to 560.

582 is between 580 and 590, but closer to 580.
So, 582 rounds to 580.

➤ To round to the *nearest hundred*

517 is between 500 and 600, but closer to 500.
So, 517 rounds to 500.

555 is between 500 and 600, but closer to 600.
So, 555 rounds to 600.

582 is between 500 and 600, but closer to 600.
So, 582 rounds to 600.

Try These

1. Round each number to the nearest ten.

   a) 241 __240__   b) 872 __870__   c) 569 __570__   d) 85 __90__

2. Round each number to the nearest hundred.

   a) 117 __100__   b) 390 __400__   c) 650 __700__   d) 901 __900__

Practice

1. Play this game with a partner.
   You will need:
   2 number cubes   2 markers of different colours

   Take turns:
   ➤ Roll the cubes and add. Move your marker that many spaces.
   ➤ If you land on a circled number, your turn is over.
     Otherwise, round the number you land on to the nearest ten and move
     your marker to that number. Consider Start to be 600.
   ➤ Keep playing until one player gets to 700.

Start	601	602	603	604	605	606	607	608	609	(610)
(620)	619	618	617	616	615	614	613	612	611
621	622	623	624	625	626	627	628	629	(630)
(640)	639	638	637	636	635	634	633	632	631
641	642	643	644	645	646	647	648	649	(650)
(660)	659	658	657	656	655	654	653	652	651
661	662	663	664	665	666	667	668	669	(670)
(680)	679	678	677	676	675	674	673	672	671
681	682	683	684	685	686	687	688	689	(690)
700	699	698	697	696	695	694	693	692	691

Stretch Your Thinking

Suppose you round to the nearest hundred.

What is the greatest number that will round to 900? __949__

What is the least number that will round to 900? __850__

Patterns in an Addition Chart

Quick Review

At Home At School

This is an addition chart.
It shows **addition facts** to $9 + 9$.

There are patterns in
the addition chart.

+	0	1	2	3	4	5	6	7	8	9
0	0	1	2	3	4	5	6	7	8	9
1	1	2	3	4	5	6	7	8	9	10
2	2	3	4	5	6	7	8	9	10	11
3	3	4	5	6	7	8	9	10	11	12
4	4	5	6	7	8	9	10	11	12	13
5	5	6	7	8	9	10	11	12	13	14
6	6	7	8	9	10	11	12	13	14	15
7	7	8	9	10	11	12	13	14	15	16
8	8	9	10	11	12	13	14	15	16	17
9	9	10	11	12	13	14	15	16	17	18

➤ When you add, order does
not matter.
$7 + 5 = 12$ $5 + 7 = 12$

➤ Adding 0 does not change
the starting number.
$7 + 0 = 7$ $9 + 0 = 9$

➤ When you add two numbers
that are the same, you add
doubles. Doubles have a
sum that is even.
$6 + 6 = 12$ $9 + 9 = 18$

There is a pattern when you look at all the ways to find a sum.

Try These

1. Use the addition chart above.
   a) List all the pairs of numbers with a sum of 16.

   9 and 7, 8 and 8, 7 and 9

   b) List all the doubles facts.

   $0 + 0 = 0$, $1 + 1 = 2$, $2 + 2 = 4$, $3 + 3 = 6$, $4 + 4 = 8$, $5 + 5 = 10$,

   $6 + 6 = 12$, $7 + 7 = 14$, $8 + 8 = 16$, $9 + 9 = 18$

1. Use the addition chart.
   a) Colour 5 sums on the chart.
   b) Write the addition fact for each sum.

   $3 + 4 = 7$

   $6 + 6 = 12$

   $6 + 9 = 15$

   $7 + 4 = 11$

   $9 + 8 = 17$

Sample Answers

+	0	1	2	3	4	5	6	7	8	9
0	0	1	2	3	4	5	6	7	8	9
1	1	2	3	4	5	6	7	8	9	10
2	2	3	4	5	6	7	8	9	10	11
3	3	4	5	6	7	8	9	10	11	12
4	4	5	6	7	8	9	10	11	12	13
5	5	6	7	8	9	10	11	12	13	14
6	6	7	8	9	10	11	12	13	14	15
7	7	8	9	10	11	12	13	14	15	16
8	8	9	10	11	12	13	14	15	16	17
9	9	10	11	12	13	14	15	16	17	18

   c) List all the pairs of numbers with a sum of 15.

   6 and 9, 7 and 8, 8 and 7, 9 and 6

2. Use the addition chart above to help you.

   a) $9 + 8 = $ __17__

   b) $4 + 8 = $ __12__

   c) $3 + 9 = $ __12__

   d) $6 + 7 = $ __13__

   e) $8 + 5 = $ __13__

   f) $6 + 8 = $ __14__

   g) $8 + 0 = $ __8__

   h) $9 + 9 = $ __18__

   i) $2 + 9 = $ __11__

Stretch Your Thinking

Vanna threw 2 darts.
Her total score was an odd number.
She did not hit the bulls eye.
Where could the 2 darts have landed?
Find as many answers as you can.

Answers must equal

an odd number from 3 to 39.

Possible answers include 9 and 8,

8 and 5, 9 and 6, 6 and 5.

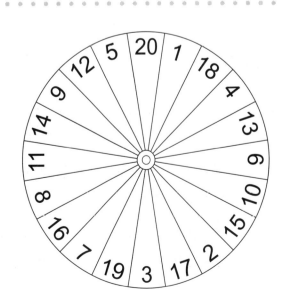

Addition Strategies

Quick Review

At Home At School

This addition chart is partly filled in.

The doubles are in the shaded diagonal. The circled diagonals show **near doubles**.

To find 7 + 8, think:

7 + 7 = 14

7 + 8 is 1 more.

So, 7 + 8 = 15

+	0	1	2	3	4	5	6	7	8	9
0	0	1								
1	1	2	3							10
2		3	4	5					10	
3			5	6	7			10		
4				7	8	9	10			
5					9	10	11			
6					10	11	12	13		
7				10			13	14	15	
8			10					15	16	17
9		10							17	18

The diagonal with the **bold** numbers shows **sums of 10**. You can use the facts for 10 to figure out other facts.

7 + 5

Think:

7 + 3, plus another 2

Make 10.

7 + 5 = 12

Try These

1. Add. Use doubles facts to help you.

a) 5 + 6 = ___11___

b) 5 + 4 = ___9___

c) 7 + 8 = ___15___

d) 8 + 9 = ___17___

e) 6 + 7 = ___13___

f) 4 + 5 = ___9___

2. Add. Use the facts for 10 to help you.

a) 9 + 5 = ___14___

b) 8 + 7 = ___15___

c) 8 + 4 = ___12___

d) 8 + 6 = ___14___

e) 5 + 8 = ___13___

f) 9 + 7 = ___16___

1. Play this game with a partner.
   You will need:
   9 small cards numbered 10 to 18 in a paper bag
   25 counters of 1 colour and 25 of another colour

   Take turns to play:
   ➤ Draw a card from the bag.
      Find 2 numbers on the game board that add up to the number on the card.
      Cover the 2 numbers with your counters.
   ➤ Put the card back in the bag.
   ➤ Play until one player cannot cover 2 numbers.

3	9	5	3	4	9	2	6
7	8	1	7	8	6	8	5
2	6	7	6	3	9	4	3
9	2	7	4	9	1	5	8
4	6	5	4	7	8	6	1
7	2	5	8	4	3	9	5

Stretch Your Thinking

Play the game again. This time, you may cover 2, 3, or 4 numbers that add to the number on the card.

Subtraction Strategies

At Home
At School

Quick Review

Here are some strategies to help you subtract.

➤ Count up through 10.

$11 - 8 =$?

Start with 8.
You need 2 more to get to 10. $8 + \mathbf{2} = 10$
You need 1 more to get to 11. $10 + \mathbf{1} = 11$
Since $\mathbf{2 + 1 = 3}$, then $11 - 8 = 3$
3 is the **difference** of 11 and 8.

➤ Count back through 10.

$12 - 4 =$?

Start with 12.
Take away 2 to get to 10: $12 - 2 = 10$
Since $4 = 2 + 2$, take away 2 more: $10 - 2 = 8$
So, $12 - 4 = 8$

Try These

1. Count up through 10 or use doubles to subtract.

   a) $13 - 8 = \underline{\quad 5 \quad}$ b) $12 - 7 = \underline{\quad 5 \quad}$ c) $11 - 9 = \underline{\quad 4 \quad}$

   d) $15 - 6 = \underline{\quad 9 \quad}$ e) $18 - 9 = \underline{\quad 9 \quad}$ f) $17 - 8 = \underline{\quad 9 \quad}$

   g) $14 - 7 = \underline{\quad 7 \quad}$ h) $15 - 8 = \underline{\quad 7 \quad}$ i) $12 - 6 = \underline{\quad 6 \quad}$

2. Count back through 10 to subtract.

   a) $12 - 5 = \underline{\quad 7 \quad}$ b) $14 - 6 = \underline{\quad 8 \quad}$ c) $11 - 8 = \underline{\quad 3 \quad}$

   d) $14 - 8 = \underline{\quad 6 \quad}$ e) $13 - 7 = \underline{\quad 6 \quad}$ f) $16 - 7 = \underline{\quad 9 \quad}$

1. Subtract to solve the riddles.
   Match each letter to its answer.

   a) Riddle 1: Where do horses go when they are sick?

$18 - 9 = $ __**9**__ (H)	$16 - 8 = $ __**8**__ (P)
$13 - 8 = $ __**5**__ (R)	$10 - 9 = $ __**1**__ (O)
$13 - 6 = $ __**7**__ (I)	$15 - 9 = $ __**6**__ (T)
$11 - 9 = $ __**2**__ (E)	$10 - 6 = $ __**4**__ (L)
$9 - 9 = $ __**0**__ (A)	$11 - 8 = $ __**3**__ (S)

T	O		T	H	E		H	O	R	S	E	-	P	I	T	A	L
6	1		6	9	2		9	1	5	3	2	-	8	7	6	0	4

   b) Riddle 2: What do elves learn in school?

$12 - 9 = $ __**3**__ (L)	$13 - 9 = $ __**4**__ (F)	$14 - 8 = $ __**6**__ (A)
$14 - 7 = $ __**7**__ (B)	$17 - 8 = $ __**9**__ (E)	$12 - 4 = $ __**8**__ (T)
$12 - 7 = $ __**5**__ (H)		

T	H	E		E	L	F	-	A	B	E	T
8	5	9		9	3	4	-	6	7	9	8

Stretch Your Thinking

Write as many subtraction facts as you can that have an answer of 8.
Sample Answer

$18 - 10 = 8$	$15 - 7 = 8$	$12 - 4 = 8$	$9 - 1 = 8$
$17 - 9 = 8$	$14 - 6 = 8$	$11 - 3 = 8$	$8 - 0 = 8$
$16 - 8 = 8$	$13 - 5 = 8$	$10 - 2 = 8$	

Related Facts

At Home
At School

Quick Review

Some number facts are **related**.
If you know one fact, you can use it to write other facts.

If you know $\qquad$ $7 + 8 = 15$
then you know $\qquad$ $8 + 7 = 15$ $\qquad$ and you know $\qquad$ $15 - 8 = 7$
$15 - 7 = 8$

If you know $\qquad$ $9 + 9 = 18$
then you know $\qquad$ $18 - 9 = 9$

A doubles fact gives us only one other fact.

Try These

1. Use each set of numbers to write a set of related facts.

   a) 6, 4, 10 $\underline{\textbf{6 + 4 = 10} \qquad \textbf{4 + 6 = 10} \qquad \textbf{10 - 6 = 4} \qquad \textbf{10 - 4 = 6}}$

   b) 5, 9, 14 $\underline{\textbf{5 + 9 = 14} \qquad \textbf{9 + 5 = 14} \qquad \textbf{14 - 9 = 5} \qquad \textbf{14 - 5 = 9}}$

   c) 7, 7, 14 $\underline{\textbf{7 + 7 = 14} \qquad \textbf{14 - 7 = 7}}$

   d) 9, 15, 6 $\underline{\textbf{9 + 6 = 15} \qquad \textbf{6 + 9 = 15} \qquad \textbf{15 - 9 = 6} \qquad \textbf{15 - 6 = 9}}$

2. Write the related facts for each given fact.

   a) $6 + 8 = 14$ $\underline{\textbf{8 + 6 = 14} \qquad \textbf{14 - 6 = 8} \qquad \textbf{14 - 8 = 6}}$

   b) $7 + 5 = 12$ $\underline{\textbf{5 + 7 = 12} \qquad \textbf{12 - 7 = 5} \qquad \textbf{12 - 5 = 7}}$

   c) $13 - 6 = 7$ $\underline{\textbf{13 - 7 = 6} \qquad \textbf{6 + 7 = 13} \qquad \textbf{7 + 6 = 13}}$

   d) $10 - 8 = 2$ $\underline{\textbf{10 - 2 = 8} \qquad \textbf{2 + 8 = 10} \qquad \textbf{8 + 2 = 10}}$

1. Play this game with a partner.
   You will need:
   2 sets of cards numbered 1 to 9
   A paper bag
   10 beans for each player

0	1	2	3	4
5	6	7	8	9

   ➤ Partners each pick a grid.
   ➤ Put the numbered cards
      in the bag and shake.
   ➤ Take turns.
      Draw 2 cards.
      Add or subtract the 2 numbers
      on the cards.
      Put a bean on your grid
      on the sum or the difference.
      If there is already a bean
      on the number, you cannot
      put another one there.

0	1	2	3	4
5	6	7	8	9

   ➤ Keep playing until one player
      has covered all the numbers
      on his or her grid.

Stretch Your Thinking . . . .

1. The numbers in a set of related facts are 9, 4, and ⬚.

   a) What could the missing number be? ____13____
      Write the related facts.

      $4 + 9 = 13$ $9 + 4 = 13$ $13 - 4 = 9$ $13 - 9 = 4$

   b) What is another possible missing number? ____5____
      Write the related facts.

      $4 + 5 = 9$ $5 + 4 = 9$ $9 - 4 = 5$ $9 - 5 = 4$

Find the Missing Number

Quick Review

To find the missing number, think about related facts.

$5 + \boxed{?} = 11$

Think:

Solve: $5 + 6 = 11$

You could think subtraction.

$\boxed{?} + 6 = 14$

Think:

$14 - 6 = 8$ Solve: $8 + 6 = 14$

Try These

1. Find each missing number.

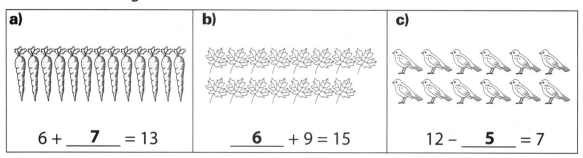

a)	b)	c)
$6 + \underline{\textbf{7}} = 13$	$\underline{\textbf{6}} + 9 = 15$	$12 - \underline{\textbf{5}} = 7$

2. Pono baked 12 muffins.
    After breakfast there were 4 muffins left.
    How many muffins were eaten at breakfast?
    Use pictures, numbers, and words to show
    your answer.

    Eight muffins were eaten at breakfast.

Sample Answer

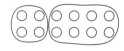

12 − 8 = 4

1. Find each missing number. Draw a picture for each.

a)	b)	c)
11 – __7__ = 4	6 + __9__ = 15	__9__ + 4 = 13
d)	e)	f)
14 – __8__ = 6	__11__ – 2 = 9	9 + __7__ = 16

2. Find each missing number.

a) 12 – __3__ = 9   b) __9__ + 7 = 16   c) __10__ – 8 = 2

d) 3 + __9__ = 12   e) 15 – __8__ = 7   f) 5 + __8__ = 13

g) 15 – __6__ = 9   h) __14__ – 5 = 9   i) __9__ + 9 = 18

3. What number do you subtract from 11 to make 9? Explain.

Sample Answer: 11 – 2 = 9. I know this because 9 + 2 = 11.

Stretch Your Thinking

1. Find the missing numbers: _____ – 8 = _____
   Show as many different ways as you can.

   Sample Answer

18 – 8 = 10	**15 – 8 = 7**	**12 – 8 = 4**	**9 – 8 = 1**
17 – 8 = 9	**14 – 8 = 6**	**11 – 8 = 3**	**8 – 8 = 0**
16 – 8 = 8	**13 – 8 = 5**	**10 – 8 = 2**	

UNIT 2
Adding and Subtracting 2-Digit Numbers

STUDENT BOOK 6 LESSON

Quick Review

➤ Find: 34 + 18
Here is one way.

Tens	Ones

Tens	Ones

Tens	Ones

Make 10.
40 + 10 + 2 = 52
34 + 18 = 52

➤ Find: 34 – 18
Here is one way.

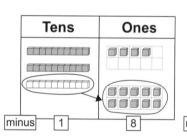

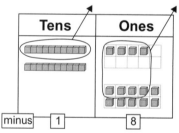

Tens	Ones

minus 1 8

minus 1 8

There are not enough ones to take away 8. Trade 1 ten for 10 ones.

Take away 8 ones. Take away 1 ten.

34 – 18 = 16

Try These

1. Add.

   a) 35 + 22 = __57__   b) 28 + 41 = __69__   c) 37 + 53 = __90__

2. Subtract.

   a) 46 – 13 = __33__   b) 57 – 35 = __22__   c) 52 – 26 = __26__

38

1. Add.

   a) $45 + 13 =$ ___**58**___   b) $67 + 19 =$ ___**86**___   c) $49 + 32 =$ ___**81**___

2. Subtract.

   a) $86 - 42 =$ ___**44**___   b) $51 - 32 =$ ___**19**___   c) $25 - 17 =$ ___**8**___

3. Add or subtract.

   a)
   $$\begin{array}{r} 83 \\ -\ 28 \\ \hline \mathbf{55} \end{array}$$
   b)
   $$\begin{array}{r} 33 \\ +\ 48 \\ \hline \mathbf{81} \end{array}$$
   c)
   $$\begin{array}{r} 79 \\ +\ 18 \\ \hline \mathbf{97} \end{array}$$
   d)
   $$\begin{array}{r} 39 \\ -\ 14 \\ \hline \mathbf{25} \end{array}$$

   e)
   $$\begin{array}{r} 51 \\ +\ 49 \\ \hline \mathbf{100} \end{array}$$
   f)
   $$\begin{array}{r} 71 \\ -\ 24 \\ \hline \mathbf{47} \end{array}$$
   g)
   $$\begin{array}{r} 53 \\ -\ 39 \\ \hline \mathbf{14} \end{array}$$
   h)
   $$\begin{array}{r} 17 \\ +\ 64 \\ \hline \mathbf{81} \end{array}$$

4. Keaka found 62 chestnuts. Leslie found 39 chestnuts.
   How many more chestnuts than Leslie did Keaka find?

   ___**62 – 39 = 23; Keaka found 23 more chestnuts than Leslie.**___

Stretch Your Thinking

Suppose you could choose 2 boxes of markers.
Find all the possible pairs.

15 Markers   **24** Markers   **36** Markers   **48** Markers

Write a number sentence to show how many markers are in each pair.

15 + 24 = 39	**15 + 36 = 51**	**15 + 48 = 63**
24 + 36 = 60	**24 + 48 = 72**	**36 + 48 = 84**

Using Mental Math to Add

Quick Review

When you add in your head, you do **mental math**.

Jake bought 28 guppies and 24 goldfish.
How many fish did Jake buy altogether?

Here are some mental math strategies to find 28 + 24.

➤ Add on tens,
   then add on ones.

➤ Take from one
   to give to the other.

Think: 28 + 20 + 4

28 + 20 = 48
48 + 4 = 52

Think: 28 + 2 + 22   28 + 24

28 + 2 = 30   ↓   ↓
30 + 22 = 52   30   22

Jake bought 52 fish.

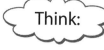

Try These

Use mental math.

1. Add.

   a) 46 + 28 = __**74**__   **b)** 18 + 24 = __**42**__   **c)** 55 + 38 = __**93**__

   d) 39 + 52 = __**91**__   **e)** 36 + 19 = __**55**__   **f)** 47 + 29 = __**76**__

2. Add. Write about the patterns you see.

   a) 36 + 10 = __**46**__ , 36 + 20 = __**56**__ , 36 + 30 = __**66**__ , 36 + 40 = __**76**__

   __The second number goes up by 10. The sums increase by 10.__

   b) 30 + 16 = __**46**__ , 30 + 26 = __**56**__ , 30 + 36 = __**66**__ , 30 + 46 = __**76**__

   __The second number goes up by 10. The sums increase by 10.__

1. Use mental math to add.

   a) 49 + 23 = __**72**__   b) 51 + 37 = __**88**__   c) 64 + 19 = __**83**__

   d) 31 + 49 = __**80**__   e) 17 + 39 = __**56**__   f) 54 + 23 = __**77**__

2. Use mental math. Find out how many seashells you would have if you bought one tub each of:

   a) sand dollars and cowries __**73**__

   b) oysters and pukas __**48**__

   c) pukas and sand dollars __**57**__

   d) pukas and cowries __**80**__

   e) oysters and cowries __**64**__

   f) sand dollars and oysters __**41**__

3. Sanjay has 27 seahorses and 26 sea urchins in his salt-water tank.

   How many sea creatures is that? __**53**__

4. Marta had 41 red buttons and 57 silver buttons.

   How many buttons is that? __**98**__

Stretch Your Thinking

1. a) Use mental math to add: 24 + 37 + 26 = __**87**__

   b) Describe the strategy you used.
   Sample Answer

   __I added 20 + 30 + 20 = 70.__

   __Then I added 4 + 7 + 6 = 17.__

   __Then I added 70 + 17 = 87.__

Using Mental Math to Subtract

At Home At School

Quick Review

You can use mental math to subtract.

Hannah collected 73 acorns.
She gave 36 acorns to Corey.
How many acorns did Hannah have left?

Here are some mental math strategies to find 73 – 36.

➤ Take away tens, then
take away ones.

Think:

$73 - 30 = 43$
Count back: $43 - 6 = 37$

➤ Add to match the
ones, then subtract.

Think:

Add 3 to 73 to make 76.
$76 - 36 = 40$
Take away the 3 you added.
$40 - 3 = 37$

Hannah had 37 acorns left.

Try These

Use mental math.

1. Subtract.

   a) $72 - 29 =$ __**43**__   **b)** $68 - 39 =$ __**29**__   **c)** $53 - 31 =$ __**22**__

   a) $43 - 27 =$ __**16**__   **b)** $38 - 19 =$ __**19**__   **c)** $86 - 27 =$ __**59**__

2. Subtract.

	a)	**b)**	**c)**	**d)**
	51	92	47	63
	− 36	− 64	− 38	− 27
	15	**28**	**9**	**36**

1. Play this game with a partner.
You will need:
10 counters each
a calculator

Take turns.
➤ Cover 2 numbers on the grid with counters.
➤ Use mental math to subtract.
➤ Record your answer in the chart.
➤ Keep playing until all the numbers have been used.
➤ Use the calculator to find your total score.
➤ The player with the greater total wins.

82	31	68	55	17
27	75	99	43	60
14	57	32	89	77
65	24	90	45	27

Player 1	Player 2
Total:	Total:

Stretch Your Thinking

Describe two ways to use mental math to find 82 – 47.
Sample Answer

1. First take away tens. 82 – 40 = 42

   Then take away ones. 42 – 7 = 35

2. First add 5 onto 82 so the ones match the ones in 47. 87 – 47 = 40

   Then take 5 away. 40 – 5 = 35

Estimating Sums and Differences

At Home
At School

Quick Review

When you do not need an exact answer, you **estimate**.

Bella has 587 silver stars and 219 gold stars.
About how many stars does Bella have?
About how many more silver stars than gold stars are there?

Estimate: 587 + 219 and 587 − 219
Here are two ways to estimate.

➤ **Rounding First**

Round to the nearest 100.

587 ⟶ 600
219 ⟶ 200

600 + 200 = 800
and
600 − 200 = 400

Bella has about 800 stars.
There are about 400 more silver stars than gold stars.

These estimates are closer.

➤ **Front-End Estimation**

Use the digits in the hundreds place. Ignore the other digits.

587 ⟶ 500
219 ⟶ 200

500 + 200 = 700
and
500 − 200 = 300

Bella has about 700 stars.
There are about 300 more silver stars than gold stars.

Try These

1. Circle the better estimate for each sum or difference.

a)
$$516$$
$$+\ 194$$
600 or ⟨700⟩

b)
$$401$$
$$+\ 394$$
700 or ⟨800⟩

c)
$$385$$
$$-\ 126$$
200 or ⟨300⟩

d)
$$613$$
$$-\ 209$$
300 or ⟨400⟩

1. Circle the two numbers that will give the sum closest to:

    a) 400:   94   (215)   85   (197)   391

    b) 700:   (229)   384   601   (501)   590

    c) 900:   146   (204)   895   (672)   284

Sample Answers

2. Estimate each sum and difference.

Problem	Estimate
426 + 109	**500**
384 + 229	**600**
114 + 201	**300**
777 + 95	**900**
231 + 286	**500**

Problem	Estimate
195 – 101	**100**
362 – 293	**100**
685 – 315	**400**
437 – 190	**200**
511 – 79	**400**

3. The estimated sum of 2 numbers is 400.
   What might the 2 numbers be?
   Give 2 different answers.

   198 and 207; 316 and 95

4. The estimated difference of 2 numbers is 100.
   What might the 2 numbers be?
   Give 2 different answers.

   485 and 399; 714 and 602

Stretch Your Thinking

Circle the two numbers that will give the difference closest to 200.

12   189   463   (701)   (908)   251

UNIT 2

STUDENT BOOK 11 LESSON

Adding 3-Digit Numbers

Quick Review

The bakery shop made 158 blueberry muffins and 156 bran muffins. How many muffins is that?

Here is one way to add 158 and 156.

➤ Use Base Ten Blocks to make 158 and 156.

Hundreds	Tens	Ones

➤ Combine the groups. Make 10.

Hundreds	Tens	Ones

➤ Trade 10 ones for 1 ten.

Hundreds	Tens	Ones

➤ Trade 10 tens for 1 hundred.

Hundreds	Tens	Ones

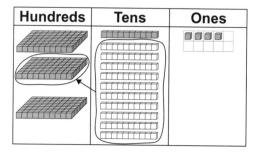

$158 + 156 = 300 + 10 + 4 = 314$
There are 314 muffins.

Try These

1. Add.

a)	143	b)	276	c)	567	d)	476
	+ 312		+ 314		+ 272		+ 335
	455		**590**		**839**		**811**

46

1. Play this game with a partner.

You will need:

1 number cube

Take turns:

➤ Roll the number cube.
Record the digit rolled in one of the boxes in your partner's
first addition problem.
Then, your partner rolls and records the digit in one of your boxes.

➤ After 6 turns each, add the numbers in your own problem.
The player with the greater sum wins.

➤ Repeat the game with the other problems.

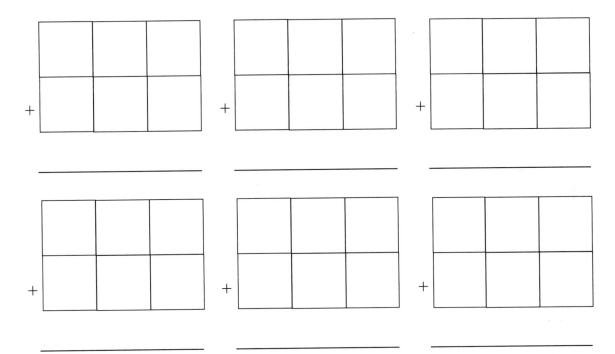

Stretch Your Thinking

The sum of 2 numbers is 427.
What might the numbers be?
Find 3 different answers.

Sample Answer: 200 and 227; 150 and 277; 307 and 120

Subtracting 3-Digit Numbers

Quick Review

Mina has 185 pennies. Wayne has 324 pennies.
How many more pennies does Wayne have?

Here is one way to find 324 − 185.

➤ To subtract 185, take away
1 hundred 8 tens 5 ones.

Hundreds	Tens	Ones

minus — 1 — 8 — 5

➤ There are not enough ones.
Trade 1 ten for 10 ones.

Hundreds	Tens	Ones

minus — 1 — 8 — 5

➤ There are not enough tens.
Trade 1 hundred for 10 tens.

Hundreds	Tens	Ones

minus — 1 — 8 — 5

➤ Take away 1 hundred, 8 tens,
and 5 ones.

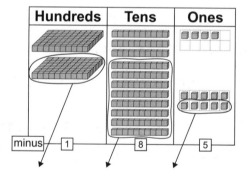

Hundreds	Tens	Ones

minus — 1 — 8 — 5

324 − 185 = 139
Wayne has 139 more pennies than Mina.

Try These

1. Subtract.

   a)   476
       − 223
       253

   b)   571
       − 348
       223

   c)   624
       − 235
       389

   d)   804
       − 521
       283

1. Subtract.

   a) 294
      − 38
      256

   b) 763
      − 521
      242

   c) 486
      − 247
      239

   d) 309
      − 142
      167

   e) 550
      − 319
      231

   f) 800
      − 289
      511

   g) 638
      − 259
      379

   h) 975
      − 487
      488

2. Use the data in the chart to answer each question.

   Stamps Collected

Name	Number of Stamps
Noah	327
Reba	241
Lily	638
Lokahi	509
Cindy	400

   a) How many more stamps did Cindy collect than Reba? **159**

   b) How many more stamps did Lily collect than Lokahi? **129**

   c) Who collected 86 more stamps than Reba? **Noah**

   d) Who collected 109 fewer stamps than Lokahi? **Cindy**

   e) What is the difference between the greatest number of stamps collected and the least number? **638 − 241 = 397**

3. Paolo and Nawel go to a campground 762 km from home. They travel 537 km by train. The rest of the trip is by bus.

   How far do they travel by bus? **225 km**

Stretch Your Thinking

Find two 3-digit numbers that subtract to leave 241.
Show your work.

Sample Answer: 597 − 241 = 356; so 597 − 356 = 241

A Standard Method for Addition

Quick Review

Vinh made two paper chains.
One chain had 216 links. The other had 379 links.
How many links is that altogether?

Find: 216 + 379
Estimate first: 200 + 400 = 600

Here is one way to add.

Add the ones. Trade 10 ones for 1 ten.	Add the tens.	Add the hundreds.
$2\overset{1}{1}6$ $+\ 379$ **5**	$2\overset{1}{1}6$ $+\ 379$ **95**	$\overset{1}{2}16$ $+\ 379$ **595**
15 ones = 1 ten + 5 ones	9 tens	5 hundreds

216 + 379 = 595
Vinh's paper chains had a total of 595 links.

Try These

1. Estimate first. Then add.

   a)   35
        + 24
          59

   b)   78
        + 18
          96

   c)   561
        + 387
         948

   d)   381
        + 299
         680

   e)   200
        + 146
         346

   f)   384
        + 96
         480

   g)   84
        + 27
         111

   h)   257
        + 135
         392

Practice

1. Add.

a)
```
   48
 + 39
   87
```

b)
```
   62
 + 13
   75
```

c)
```
   87
 + 49
  136
```

d)
```
   25
 + 76
  101
```

e)
```
  276
 + 121
  397
```

f)
```
  381
 + 494
  875
```

g)
```
  609
 + 148
  757
```

h)
```
  357
 + 641
  998
```

2. Estimate first. Then, solve only the problems with a sum greater than 600.

a)
```
   49
 + 99
```

b)
```
  427
 + 198
  625
```

c)
```
  368
 + 431
  799
```

d)
```
  284
 + 379
  663
```

3. Mr. Tanaka drove 376 km on Thursday and 489 km on Friday.
   How far did Mr. Tanaka drive on the two days?

   376 + 489 = 865; Mr. Tanaka drove 865 km.

4. a) Forty-two Grade 3 children went to the zoo.
   Thirty-eight Grade 2 children went along with them.

   How many children went to the zoo? **80 children**

   b) Fifty-seven Grade 4 children joined the others for lunch at the zoo.

   How many children had lunch together at the zoo? **137 children**

Stretch Your Thinking

1. What is the greatest number you can add to 176 to get a sum less than 300?

   123

2. What is the least number you can add to 176 to get a sum greater than 300?

   125

A Standard Method for Subtraction

At Home
At School

Quick Review

There are 508 children at the Science Centre.
Two hundred and seventy-nine are boys.
How many girls are there?

Find: 508 – 279

Here is one way to subtract.
You cannot subtract the ones.
There are no tens to trade.
Go to the hundreds.

$$\begin{array}{r} 508 \\ -\ 279 \end{array}$$

Trade 1 hundred for 10 tens.	Trade 1 ten for 10 ones. Subtract the ones.	Subtract the tens. Subtract the hundreds.
$$\begin{array}{r} {}^{4\ 10}\\ \cancel{508} \\ -\ 279 \end{array}$$	$$\begin{array}{r} {}^{9}\\ {}^{4\ \cancel{10}\ 18}\\ \cancel{508} \\ -\ 279 \\ \hline 9 \end{array}$$	$$\begin{array}{r} {}^{9}\\ {}^{4\ \cancel{10}\ 18}\\ \cancel{508} \\ -\ 279 \\ \hline 229 \end{array}$$

There are 229 girls.

Try These

1. Subtract.

a)
$$\begin{array}{r} 76 \\ -\ 32 \\ \hline \mathbf{44} \end{array}$$

b)
$$\begin{array}{r} 84 \\ -\ 25 \\ \hline \mathbf{59} \end{array}$$

c)
$$\begin{array}{r} 95 \\ -\ 48 \\ \hline \mathbf{47} \end{array}$$

d)
$$\begin{array}{r} 78 \\ -\ 39 \\ \hline \mathbf{39} \end{array}$$

e)
$$\begin{array}{r} 372 \\ -\ 211 \\ \hline \mathbf{161} \end{array}$$

f)
$$\begin{array}{r} 470 \\ -\ 255 \\ \hline \mathbf{215} \end{array}$$

g)
$$\begin{array}{r} 509 \\ -\ 141 \\ \hline \mathbf{368} \end{array}$$

h)
$$\begin{array}{r} 600 \\ -\ 231 \\ \hline \mathbf{369} \end{array}$$

Practice

1. Subtract

   a)   88
       − 47
      41

   b)   72
       − 38
      34

   c)   90
       − 34
      56

   d)   75
       − 26
      49

   e)   473
       − 148
      325

   f)   827
       − 351
      476

   g)   608
       − 234
      374

   h)   800
       − 196
      604

2. Use the data in the chart.

Grade	Number of Tags
K	368
1	426
2	219
3	509

   a) How many more tags did Grade K collect than Grade 2? **149**

   b) How many more tags would Grade 2 have to collect to be even with Grade 1? **207**

   c) What is the difference between the least number and the greatest number collected? **509 − 219 = 290**

3. Jacob went to the flea market with 99¢.
   He bought a comic book for 35¢ and a toy soldier for 50¢.
   Then his uncle gave him 38¢ more.
   Does Jacob have enough money to buy another toy soldier for 50¢?
   How do you know?

   Sample Answer: Yes; Jacob spent 35¢+50¢=85¢.

   He was left with 99¢−85¢=14¢. His uncle added 38¢, so 14¢+38¢=52¢.

Stretch Your Thinking

Use these numbers: 3, 5, 6, 7, 8, 9
Arrange the numbers to make
the greatest possible difference.

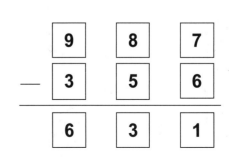

Describing Figures

Quick Review

An **attribute** is a way to describe a figure.
Here are some attributes of figures.

➤ The lengths of the sides

This figure has some
sides the same length.

We use hatch marks to
show equal lengths.

This figure has all sides
the same length.

➤ The direction of the sides

This figure has 2 pairs
of **parallel** sides.

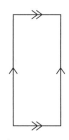

We use arrows to
show parallel lines.

This figure has
no parallel sides.

Try These

1. Tell which figures have:

   a) no parallel sides __C, F__

   b) all sides different lengths __A, C__

   c) all sides the same length __B, E__

   d) some parallel sides __A, B, D, E__

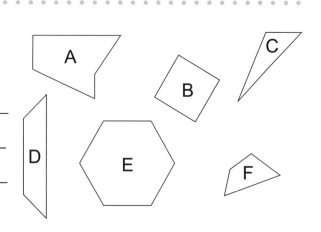

1. Find the figures below that have each of these attributes.
    Label the figures with the letters.

    A — has all sides the same length
    B — has no sides the same length
    C — has some sides the same length
    D — has 1 pair of parallel sides
    E — has 2 pairs of parallel sides
    F — has more than 2 pairs of parallel sides
    G — has no parallel sides

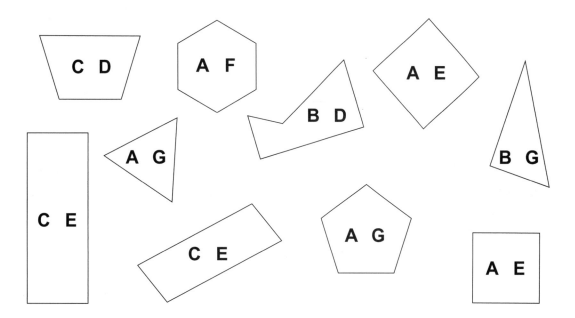

Stretch Your Thinking

Draw as many figures as you can with 2 pairs of parallel sides.
Sample Answer

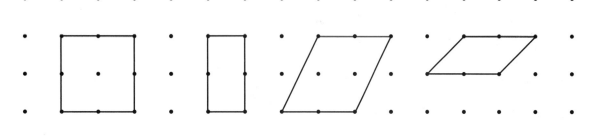

Describing Angles

Quick Review

Two sides of a figure meet at a **vertex**.
The two sides make an **angle**.
This figure has four angles.

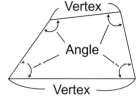

When the sides make a square corner,
the angle is a **right angle**.
These figures have right angles.

These figures have angles that
are greater than a right angle.

These figures have angles that
are less than a right angle.

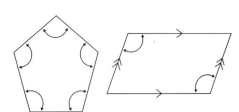

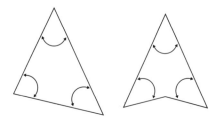

Try These

1. Tell which figures have:

   a) three right angles **B**

   b) no right angles **C, D**

   c) two angles greater than a right angle **C**

   d) one right angle **A**

   e) one angle less than a right angle **A, B, C, D**

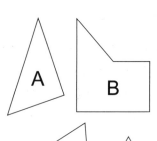

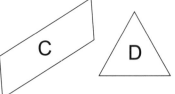

1. Use the dot paper below.
   Draw a figure that fits each description.
   Label each figure with its letter.

   A — has four right angles
   B — has only one right angle
   C — has only two right angles
   D — has no right angles
   E — has at least one angle greater than a right angle
   F — has at least one angle less than a right angle

   Sample Answer

   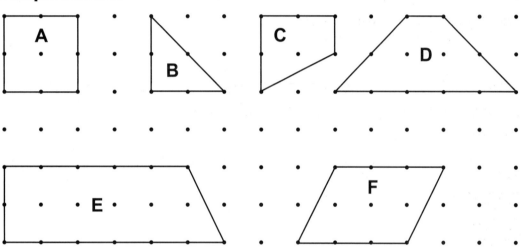

Stretch Your Thinking

Draw a figure on the dot paper.
Your figure should have:
- three right angles
- one angle less than a right angle
- one angle greater than a right angle

Sample Answer

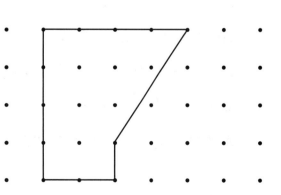

Naming Figures

At Home At School

Quick Review

➤ Here are some figures with 4 sides.
A **trapezoid** has 2 parallel sides.

A **parallelogram** has
2 pairs of parallel sides.

A **rhombus** is a parallelogram
with all sides equal.

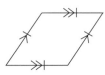

➤ Here are some figures with more than 4 sides.
A **pentagon** has 5 sides.   A **hexagon** has 6 sides.

          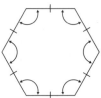

➤ A **regular figure** has all sides equal and all angles equal.
This is a regular pentagon.   This is a regular hexagon.

   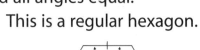

Try These

1. Use the figures. Find:

   a) a trapezoid **B**

   b) a rhombus **C**

   c) a pentagon **A**

 A    B   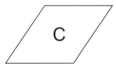 C

Practice

1. Play this game with a partner.
   You will need:
   16 beans
   a paper clip and pencil for the spinner

   ➤ Put 2 beans in each box on the game board.
   ➤ Take turns spinning the spinner. Name the figure.
   ➤ Take 1 bean from the box with the figure's name.
   If you spin a figure with no beans left in its box, you lose a turn.
   ➤ Keep playing until all the beans are gone.
   The player with the most beans is the winner.

parallelogram	triangle
trapezoid	rectangle
square	hexagon
pentagon	rhombus

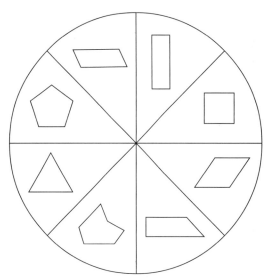

Stretch Your Thinking

Draw a figure with 8 sides.
Write about the figure.
Tell about its sides and angles.

Sample Answer

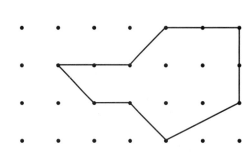

Sample Answer: My figure has

8 angles. One is a right angle.

Six are greater than a right angle.

One is less than a right angle.

One pair of sides is parallel.

Sorting Figures

Quick Review

You can use this **Venn diagram** to sort.

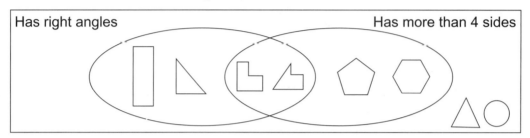

The figures in the left loop have right angles.
The figures in the right loop have more than 4 sides.
The figures in the middle have right angles *and* more than 4 sides.
The figures outside the loops have no right angles *and* they do not have more than 4 sides.

The sorting rule is:
Figures with right angles and figures with more than 4 sides

Try These

1. a) Look at these figures. Choose 2 attributes.
   Sort the figures in the Venn diagram.

Sample Answer

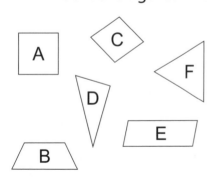
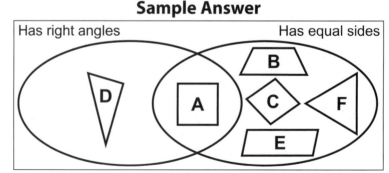

   b) Write the sorting rule.

   Figures with right angles and figures with equal sides

Sample Answers

1. Use the Venn diagram below.
   Sketch 2 different figures in each loop.
   Sketch 1 figure in the middle and 1 figure outside the loops.

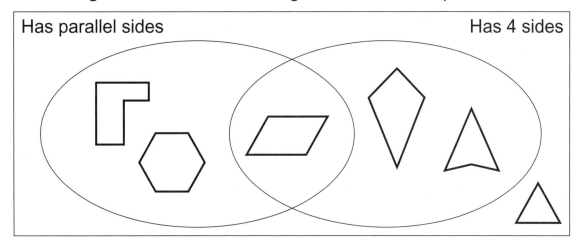

2. **a)** Draw a figure that belongs to this set.

   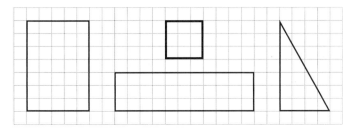

   b) How do you know this figure belongs?

   It has one or more right angles.

Stretch Your Thinking

Sample Answers

1. **a)** Choose an attribute. Sketch 4 figures with that attribute.

   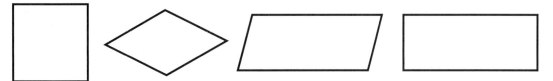

   b) Name the attribute. **2 pairs of parallel sides**

Congruent Figures

At Home
At School

Quick Review

Two figures that have the same shape and size are **congruent**.

➤ Congruent figures have equal matching sides and equal matching angles.

These trapezoids are congruent.

These triangles are *not* congruent.

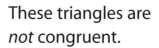

➤ To show two figures are congruent, place one figure on top of the other. If they coincide, they are congruent.

➤ You may need to flip or turn the figures to show they are congruent. If you cannot move the figures:
Trace one figure. Place it on top of the other figure.

If the tracing coincides with the other figure, the figures are congruent.

Try These

1. Find pairs of congruent figures. Colour each pair a different colour.
Sample Answer

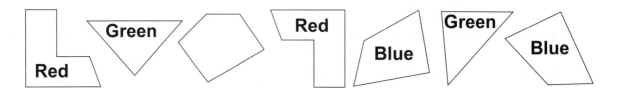
Red · · · Green · · · Red · · · Blue · · · Green · · · Blue

1. Use the dot paper below.

Draw 4 pairs of congruent figures.

Don't put the congruent figures close to each other.

Flip or turn some of the figures.

Label each figure with a different letter.

When you are done, ask a partner to find the congruent pairs.

Sample Answer

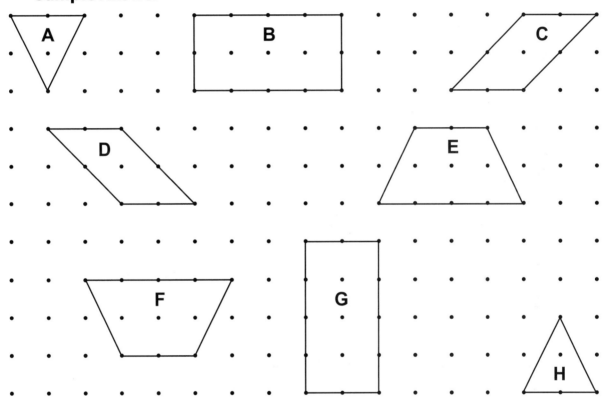

Congruent pairs: **A and H, B and G, C and D, E and F**

Stretch Your Thinking

How many congruent regular hexagons can you find in this pattern?

There are 18 regular small hexagons

and 4 regular large hexagons.

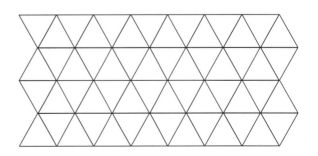

Making Pictures with Figures

Quick Review

Many figures may be made from other figures.

With 3 congruent rhombuses, you can make a hexagon.

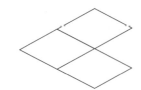

With 2 congruent trapezoids, you can make a parallelogram.

When a regular hexagon and a triangle have all sides the same length, you can make a pentagon.

Try These

1. Make each figure. Use 2 different Pattern Blocks.
   Sketch each figure you made.
   Sample Answers

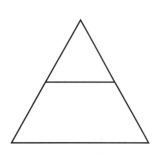 

Triangle   Pentagon   Parallelogram   Trapezoid

Practice

1. Use 4 congruent squares cut from coloured paper.
   The squares should have 4-cm sides.

   Make each square into 4 triangles
   by cutting along the diagonals like this:
   Use each set of 4 triangles
   to make a different figure.
   Equal sides of the triangles must touch.
   Glue the figures you make in this space.
   Print the name of each figure.

Sample Answer

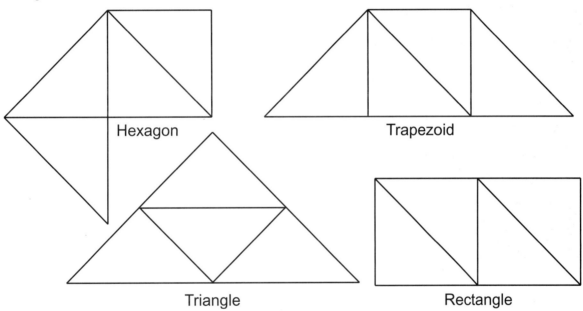

Hexagon

Trapezoid

Triangle

Rectangle

Stretch Your Thinking

Use the grid paper.
Draw as many figures as you can using 4 congruent squares.
Sample Answer

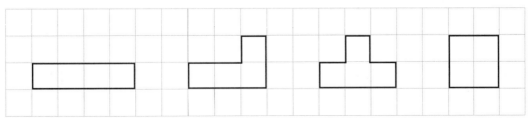

Identifying Prisms and Pyramids

Quick Review

➤ A pyramid has 1 **base**. The base is a **face**.
The shape of the base tells the name of the pyramid.
A pyramid also has triangular faces.

This is a pentagonal pyramid.
It has 6 faces:
1 pentagon
5 triangles

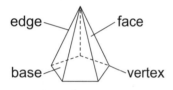

This is a triangular pyramid.
It has 4 faces:
4 triangles

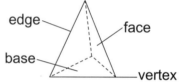

➤ A prism has 2 congruent bases.
The shape of the bases tells the name of the prism.
A prism also has rectangular faces.

This is a hexagonal prism.
It has 8 faces:
2 hexagons
6 rectangles

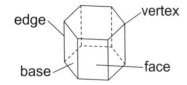

This is a rectangular prism.
It has 6 faces:
6 rectangles

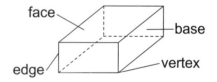

Try These

1. Colour the pyramids blue. Colour the prisms red.

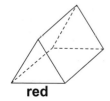

red

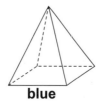

blue

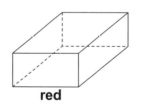

red

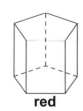

red

blue

1. Complete the chart.

Solid	Name	Number of Faces	Number of Edges	Number of Vertices
	triangular prism	5	9	6
	hexagonal pyramid	7	12	7
	pentagonal prism	7	15	10
	square pyramid	5	8	5

2. Name a pyramid and a prism that have each face.

a)    <u>hexagonal pyramid, hexagonal prism</u>

b) 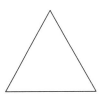   <u>Sample Answer: triangular pyramid,</u>

<u>triangular prism</u>

Stretch Your Thinking

This is an octagon. Draw the faces of an octagonal prism.

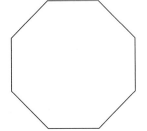

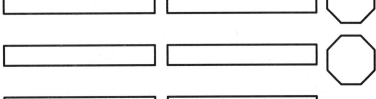

Sorting Solids

Quick Review

At Home At School

Here are some solids you know.

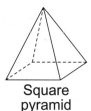
Square
pyramid

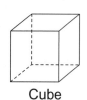
Cube

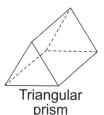
Triangular
prism

Pentagonal
prism

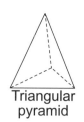
Triangular
pyramid

One way to sort these solids is in a Venn diagram, with these attributes:

• Has triangular faces

• Has all faces congruent

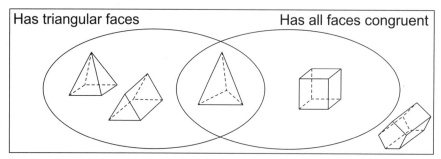

The pentagonal prism has neither of these attributes.
It is placed outside the loops.

 The sorting rule is:
Solids with triangular faces and solids with all faces congruent

Try These

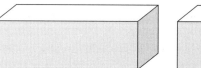

How are these 2 solids the same? Different?

Both have 8 vertices, 12 edges, and 6

faces. The cube has all faces congruent.

The prism has some congruent faces.

Practice

1. Sort these solids in the Venn diagram below.

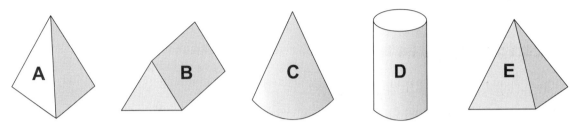

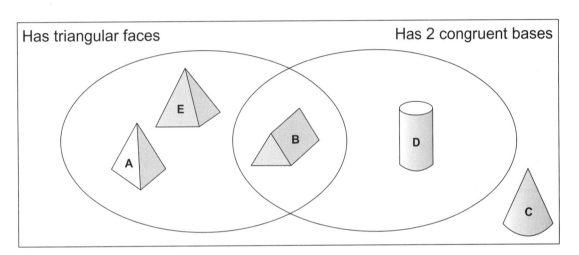

2. Write the names of one or more solids that could answer each riddle.
Sample Answers

a) I have 12 edges. **rectangular prism, hexagonal pyramid**

b) I have 6 vertices. **pentagonal pyramid, triangular prism**

c) I have 4 faces. **triangular pyramid**

Stretch Your Thinking

Find 2 different boxes or containers.
Tell how they are different and how they are the same.

Sample Answer: One box has 8 vertices, 12 edges, and 6 faces.

The other box has 5 faces, 9 edges, and 6 vertices.

One is a cube and the other is a triangular prism.

Making Models from Figures

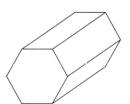

Quick Review

➤ You can use different figures to build models of solids.

This box was made from 2 hexagons and 6 rectangles.

➤ You can make a model of a solid from one cutout.
The cutout is called a **net.**

This is a net for a square pyramid. It shows all the faces of the square pyramid joined in 1 piece.

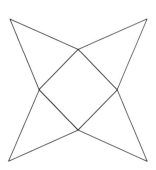

The net can be folded to make a model of a square pyramid.

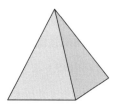

Try These

1. Which solid will each net fold into?

a)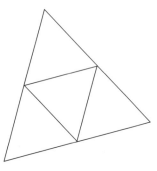

b)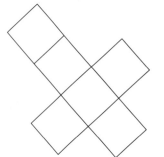

c)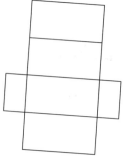

triangular pyramid_____   cube_____   rectangular prism_____

Practice

1. Name the solid you could build with each set of figures.

a)

<u>**rectangular prism**</u>

b)

<u>**hexagonal pyramid**</u>

c)

<u>**pentagonal pyramid**</u>

2. Draw the figures you would need to build a model of each solid.

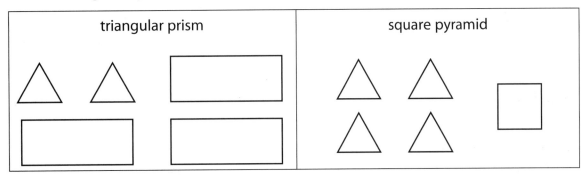

| triangular prism | square pyramid |

Stretch Your Thinking

Draw a net for this solid.

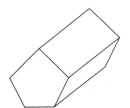

Sample Answer

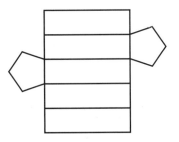

Making a Structure from Solids

Quick Review

You can build a structure from solids.

Here is a playhouse Roger built.

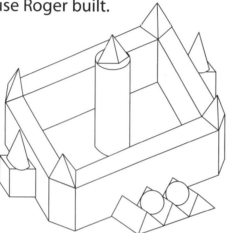

Here is a drawing Roger made of his playhouse.

Try These

1. **a)** Name the prisms you can see in Roger's playhouse.

   rectangular prism, triangular prism

   b) Name the pyramids.

   square pyramid, triangular pyramid

2. Which other solids did Roger use?

   cones, spheres, and a cylinder

Sample Answer

1. **a)** Find a picture of a structure in a newspaper or magazine.
Paste the picture below.

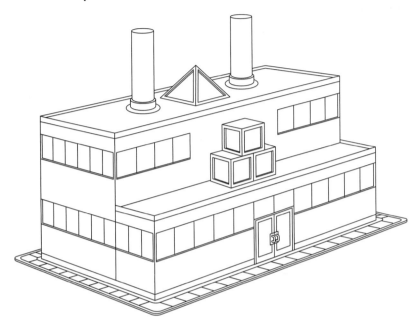

b) Describe the structure. Name the solids you can see.

The main part of the building is a rectangular prism.

On top of the rectangular prism there are two cylinders and

a square pyramid.

On top of the lower rectangular prism there are 3 cubes.

Stretch Your Thinking ·

1. Use boxes and packages. Make a tall structure.
Sketch your structure. Explain how you built it.

Sample Answer: I made my structure

with 6 boxes. I piled one box on top of

the other to build a tall tower.

I placed the widest boxes on the bottom.

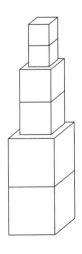

UNIT 4

STUDENT BOOK **1** LESSON

Relating Multiplication and Addition

At Home At School

Quick Review

Equal groups have the same number of things in each group. When you have equal groups, you can add or multiply to find how many altogether.

Here are 5 towers of Snap Cubes. There are 4 cubes in each tower.

To find how many Snap Cubes:

➤ You can add: 4 + 4 + 4 + 4 + 4 = 20

> **Think:** 5 groups of 4 added = 20

This is an addition sentence.

➤ You can multiply: 5 × 4 = 20

> **Think:** 5 groups of 4 = 20

This is a multiplication sentence.

Another way to say this is "5 **times** 4 equals 20."

Try These

Write an addition sentence and a multiplication sentence for each picture.

1.

2 + 2 + 2 = 6   3 × 2 = 6

2.

3 + 3 + 3 + 3 = 12   4 × 3 = 12

3.

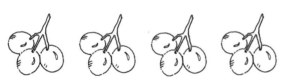

6 + 6 = 12   2 × 6 = 12

1. Write an addition sentence for each multiplication sentence.

   a) $3 \times 3 = 9$ **3 + 3 + 3 = 9**

   b) $2 \times 8 = 16$ **8 + 8 = 16**

   c) $3 \times 6 = 18$ **6 + 6 + 6 = 18**

   d) $4 \times 2 = 8$ **2 + 2 + 2 + 2 = 8**

2. Write a multiplication sentence for each addition sentence.

   a) $2 + 2 + 2 + 2 + 2 = 10$ **5 × 2 = 10**

   b) $1 + 1 + 1 = 3$ **3 × 1 = 3**

   c) $5 + 5 + 5 = 15$ **3 × 5 = 15**

   d) $7 + 7 = 14$ **2 × 7 = 14**

3. Draw a picture for each multiplication sentence.
   Then write an addition sentence.
   Sample Answers

a) $3 \times 4 = 12$	b) $2 \times 8 = 16$	c) $4 \times 6 = 24$
4 + 4 + 4 = 12	**8 + 8 = 16**	**6 + 6 + 6 + 6 = 24**

Write an addition sentence and a multiplication sentence
to find how many eggs in 4 dozen.

12 + 12 + 12 + 12 = 48   **4 × 12 = 48**

Using Arrays to Multiply

At Home
At School

Quick Review

An **array** shows objects arranged in equal rows.

➤ Here are 2 arrays that show 12 counters.
You can use an addition sentence and a multiplication sentence to tell how many counters.

4 rows of 3   ○ ○ ○   3 rows of 4   ○ ○ ○ ○
$3 + 3 + 3 + 3 = 12$ ○ ○ ○ $4 + 4 + 4 = 12$ ○ ○ ○ ○
4 groups of 3   ○ ○ ○   3 groups of 4   ○ ○ ○ ○
$4 \times 3 = 12$ ○ ○ ○ $3 \times 4 = 12$

➤ You can use an array to multiply.
To find 2×3, make an array of 2 rows of 3.

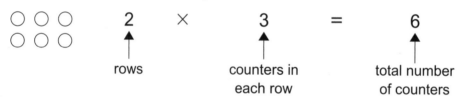

○ ○ ○   2   ×   3   =   6
○ ○ ○   ↑   ↑   ↑

          rows   counters in   total number
                  each row   of counters

In a multiplication sentence,
the numbers you multiply
are **factors**.   2   ×   3   =   6
The answer is the **product**.   factor   factor   product

Try These ·

Write a multiplication sentence for each array.

1.

2.

3.

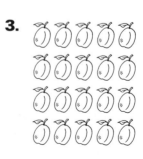

    $3 \times 5 = 15$ $2 \times 4 = 8$ $4 \times 5 = 20$

Practice

1. Draw an array to find each product.

a) $4 \times 4 =$ __16__	**b)** $3 \times 7 =$ __21__	**c)** $6 \times 2 =$ __12__
d) $3 \times 6 =$ __18__	**e)** $2 \times 5 =$ __10__	**f)** $5 \times 3 =$ __15__

2. Use counters. Make an array to find each product.

a) $7 \times 2 =$ __14__   **b)** $6 \times 4 =$ __24__   **c)** $7 \times 3 =$ __21__

d) $3 \times 3 =$ __9__   **e)** $5 \times 7 =$ __35__   **f)** $5 \times 5 =$ __25__

3. There are 6 rows of marchers in the band.
   There are 7 marchers in each row.

   How many marchers are there in all? **$6 \times 7 = 42$**

Stretch Your Thinking

Find as many ways of arranging 24 counters in equal rows as you can.
Write a multiplication sentence for each way.

$1 \times 24 = 24$; $2 \times 12 = 24$; $3 \times 8 = 24$; $4 \times 6 = 24$; $6 \times 4 = 24$;

$8 \times 3 = 24$; $12 \times 2 = 24$; $24 \times 1 = 24$

Multiplying by 2 and by 5

Quick Review

➤ Multiply 6×2.
  - One way to multiply is to skip count on a number line.

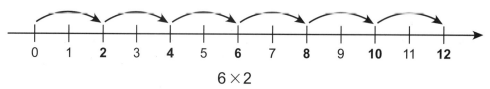

6×2

  - Another way to multiply 6 by 2 is to use doubles.
    Double 6 is 12.

➤ Multiply 4×5.
  - One way to multiply
    4 by 5 is to skip count
    on a hundred chart.

1	2	3	4	⑤	6	7	8	9	⑩
11	12	13	14	⑮	16	17	18	19	⑳

Try These

1. Write a multiplication sentence for each picture.

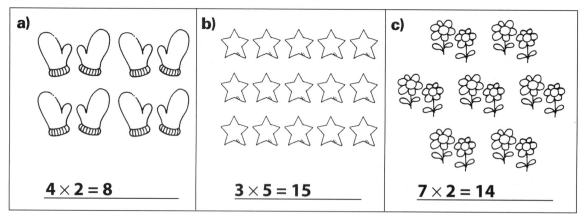

   a)
   $\underline{4 \times 2 = 8}$

   b)
   $\underline{3 \times 5 = 15}$

   c)
   $\underline{7 \times 2 = 14}$

2. Use a hundred chart to multiply.

   a) $6 \times 5 = \underline{\quad 30 \quad}$

   b) $2 \times 5 = \underline{\quad 10 \quad}$

   c) $7 \times 5 = \underline{\quad 35 \quad}$

Practice

1. Use each number line to multiply.

a) $5 \times 2 =$ **10**
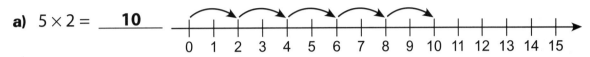

b) $2 \times 6 =$ **12**
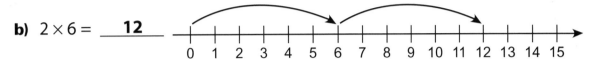

c) $5 \times 1 =$ **5**
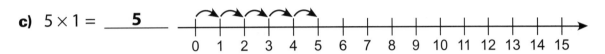

2. Multiply to find the answer to the riddle.
   Match each letter to its answer in the box below.
   Some letters are not used.

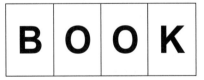

What do you call a funny book about eggs?

$4 \times 5 =$ **20** (X) $5 \times 3 =$ **15** (K) $6 \times 2 =$ **12** (B)

$2 \times 3 =$ **6** (R) $6 \times 5 =$ **30** (S) $2 \times 5 =$ **10** (L)

$1 \times 5 =$ **5** (A) $2 \times 4 =$ **8** (Y) $2 \times 1 =$ **2** (H)

$2 \times 2 =$ **4** (P) $2 \times 7 =$ **14** (N) $7 \times 5 =$ **35** (O)

A	Y	O	L	K		B	O	O	K
5	8	35	10	15		12	35	35	15

Stretch Your Thinking

There were 6 bicycles, 5 tricycles, and 4 wagons in the playground.
How many wheels were there?

Sample Answer: $6 \times 2 = 12$; $5 \times 3 = 15$; $4 \times 4 = 16$

$12 + 15 + 16 = 43$; There are 43 wheels.

Multiplying by 10

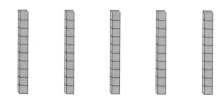

Quick Review

Here are 3 ways to multiply by 10.

➤ Use Base Ten Blocks and skip count.

Count: 10   20   30   40   50

$5 \times 10 = 50$

➤ Skip count on a number line.

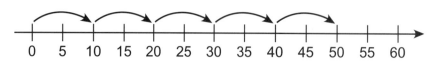

$5 \times 10 = 50$

➤ Use patterns and place value.

The number multiplied by 10 is the same as the tens digit of the product.

$1 \times 10 = 10$
$2 \times 10 = 20$
$3 \times 10 = 30$
$4 \times 10 = 40$
$\mathbf{5} \times 10 = \mathbf{5}0$

The ones digit of the product is always 0.

Try These

Multiply.

1. a) $3 \times 10 = \underline{\quad 30 \quad}$ b) $7 \times 10 = \underline{\quad 70 \quad}$ c) $10 \times 2 = \underline{\quad 20 \quad}$

   d) $8 \times 10 = \underline{\quad 80 \quad}$ e) $10 \times 1 = \underline{\quad 10 \quad}$ f) $6 \times 10 = \underline{\quad 60 \quad}$

   g) $4 \times 10 = \underline{\quad 40 \quad}$ h) $10 \times 7 = \underline{\quad 70 \quad}$ i) $10 \times 5 = \underline{\quad 50 \quad}$

Practice

1. Write a multiplication sentence for each picture.

a)

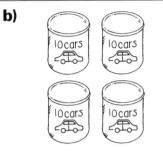

$7 \times 10 = 70$

b)

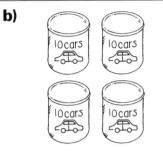

$4 \times 10 = 40$

c)

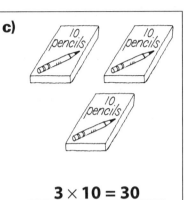

$3 \times 10 = 30$

2. Multiply.

a) $2 \times 10 = \underline{20}$

$10 \times 2 = \underline{20}$

b) $5 \times 10 = \underline{50}$

$10 \times 5 = \underline{50}$

c) $6 \times 10 = \underline{60}$

$10 \times 6 = \underline{60}$

3. Use each number line to find the product.

a)

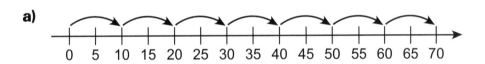

$7 \times 10 = \underline{\textbf{70}}$

b)

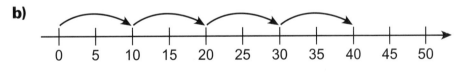

$4 \times 10 = \underline{\textbf{40}}$

Stretch Your Thinking

Danielle bought 4 packages of red pens and 6 packages of blue pens. How many pens did Danielle buy?

Sample Answer: $4 \times 10 = 40$; $6 \times 10 = 60$

$40 + 60 = 100$; Danielle bought 100 pens.

Multiplying by 1 and by 0

At Home
At School

Quick Review

Think: 5 groups of 1 is 5×1.

5	$\times$	1	$=$	5
bowls		fish		fish in all

When 1 is a factor, the product is always the other factor.

Also, $1 \times 5 = 5$

Think: 7 groups of 0 is 7×0.

7	$\times$	0	$=$	0
bowls		fish		fish in all

When 0 is a factor, the product is always 0.

Also, $0 \times 7 = 0$

Try These . . . .

Multiply.

1. a) $6 \times 1 = \underline{\quad 6 \quad}$ b) $7 \times 1 = \underline{\quad 7 \quad}$ c) $4 \times 1 = \underline{\quad 4 \quad}$

      $1 \times 6 = \underline{\quad 6 \quad}$ $1 \times 7 = \underline{\quad 7 \quad}$ $1 \times 4 = \underline{\quad 4 \quad}$

2. a) $6 \times 0 = \underline{\quad 0 \quad}$ b) $3 \times 0 = \underline{\quad 0 \quad}$ c) $2 \times 0 = \underline{\quad 0 \quad}$

      $0 \times 6 = \underline{\quad 0 \quad}$ $0 \times 3 = \underline{\quad 0 \quad}$ $0 \times 2 = \underline{\quad 0 \quad}$

1. Find each product.

a) $1 \times 4 =$ __4__   **b)** $0 \times 0 =$ __0__   **c)** $0 \times 7 =$ __0__

d) $5 \times 1 =$ __5__   **e)** $6 \times 0 =$ __0__   **f)** $1 \times 6 =$ __6__

g) $0 \times 4 =$ __0__   **h)** $7 \times 1 =$ __7__   **i)** $1 \times 1 =$ __1__

2. Find each missing number.

a) $4 \times$ __0__ $= 0$ **b)** __1__ $\times 6 = 6$ **c)** $7 \times$ __0__ $= 0$

d) __1__ $\times 1 = 1$ **e)** __0__ $\times 5 = 0$ **f)** __1__ $\times 4 = 4$

g) $1 \times$ __0__ $= 0$ **h)** __3__ $\times 1 = 3$ **i)** $2 \times$ __1__ $= 2$

3. Write $+$ or $\times$.

a) 5 __$\times$__ $1 = 5$ **b)** 1 __$\times$__ $1 = 1$ **c)** 6 __$\times$__ $0 = 0$

d) 0 __$+$__ $3 = 3$ **e)** 4 __$+$__ $1 = 5$ **f)** 0 __$\times$__ $2 = 0$

g) 1 __$\times$__ $4 = 4$ **h)** 1 __$+$__ $1 = 2$ **i)** 7 __$+$__ $0 = 7$

4. Rico has 1 nickel, 5 dimes, and 0 quarters.
How much money does Rico have?
Show your work.

5¢ + 50¢ + 0¢ = 55¢; Rico has 55¢.

Which is greater, the product of your age times 0 or

the product of your age times 1? Explain.

Sample Answer: The product of my age times 0 is 0.

The product of my age times 1 is 9.

The product of my age times 1 is greater.

UNIT 4

STUDENT BOOK
6
LESSON

Using a Multiplication Chart

Quick Review

Patterns in a multiplication chart can help you multiply.

➤ The **row** and **column** for the same factor have the same numbers. Here are the row and column for factor 4. They show that:
$4 \times 3 = 12$ and $3 \times 4 = 12$
$4 \times 7 = 28$ and $7 \times 4 = 28$

x	0	1	2	3	4	5	6	7
0					0			
1					4			
2					8			
3					12			
4	0	4	8	12	16	20	24	28
5					20			
6					24			
7					28			

➤ To fill in a row or column, skip count by the first number in the row or column.

To fill in the row or column for factor 3, start at 0, then count on by 3s:
0, 3, 6, 9, 12, 15, 18, 21

x	0	1	2	3	4	5	6	7
0				0				
1				3				
2				6				
3	0	3	6	9	12	15	18	21
4				12				
5				15				
6				18				
7				21				

Try These

1. Multiply. Use the multiplication charts above to help.

b) $3 \times 3 = $ **9**
c) $3 \times 6 = $ **18**
d) $6 \times 3 = $ **18**
e) $6 \times 4 = $ **24**
f) $3 \times 7 = $ **21**
g) $4 \times 5 = $ **20**
h) $3 \times 4 = $ **12**
i) $2 \times 4 = $ **8**

84

1. Play this game with a partner.
   You will need:
   15 counters of 2 colours
   2 markers

   ➤ Player 1: Put the 2 markers on any 2 numbers in the factor box.
   Multiply the 2 numbers.
   Put one of your counters on the product on the game board.
   ➤ Player 2: Move just 1 of the markers to a new number.
   Place one of your counters on the product.
   ➤ Continue playing. Move just 1 marker on each turn.
   ➤ The first player to make a row of 3 wins.

Factor Box

28	5	30	10	4
18	42	0	35	7
21	14	20	12	6
2	24	15	3	8

0	1
2	3
4	5
6	7

Stretch Your Thinking

Ali spent 3 weeks visiting her grandma and 2 weeks visiting her aunt.
How many days is that?

Sample Answer: 3 weeks + 2 weeks = 5 weeks

$5 \times 7 = 35$; **Ali spent 35 days visiting her grandma and aunt.**

Modelling Division

Quick Review

➤ Jacob arranges his 8 marbles into groups of 2.
How many groups are there?

There are 4 groups of 2.
You write: 8   ÷   2   =   4

             ↓   ↓   ↓
     number of   number in   number of
      marbles   each group   groups

This is a division sentence.

You say, "8 divided by 2 is 4."

➤ Now Jacob arranges the marbles into 8 equal groups.
How many marbles are in each group?

There are 8 groups of 1.
You write: 8   ÷   8   =   1

             ↓   ↓   ↓
     number of   number in   number of
      marbles   each group   groups

You say, "8 divided by 8 is 1."

Try These

1. Find the number of groups.
Then write a division sentence.
Make groups of 3.

2. Find the number in each group.
Then write a division sentence.
Make 5 equal groups.

Practice

1. Find the number of groups. Then write a division sentence.

a) Make groups of 6.	**b)** Make groups of 3.	**c)** Make groups of 5.
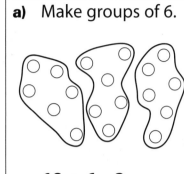	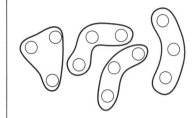	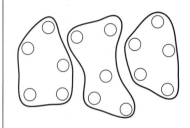
$18 \div 6 = 3$	$12 \div 3 = 4$	$15 \div 5 = 3$

2. Find the number in each group. Then write a division sentence.

a) Make 3 equal groups.	**b)** Make 6 equal groups.	**c)** Make 2 equal groups.
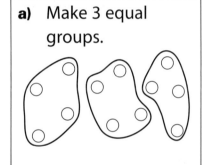		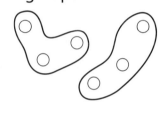
$12 \div 3 = 4$	$12 \div 6 = 2$	$6 \div 2 = 3$

3. Ira has 16 plums. He gives 4 plums to each person at his lunch table.

   How many people get plums? __**4**__

4. Sari has 15 photos. She puts 5 photos on each page.

   How many pages does Sari use? __**3**__

Stretch Your Thinking

Find as many ways to put 20 counters into equal groups as you can.
Write a division sentence for each way you find.

$20 \div 1 = 20; \ 20 \div 2 = 10; \ 20 \div 4 = 5; \ 20 \div 5 = 4; \ 20 \div 10 = 2;$

$20 \div 20 = 1$

Using Arrays to Divide

At Home
At School

Quick Review

There are 6 stools.
They will be put into equal rows.
How many stools could be in each row?

You can make an array to show each way.

2 rows of 3	3 rows of 2	1 row of 6	6 rows of 1

$6 \div 2 = 3$
2 rows of
3 stools

$6 \div 3 = 2$
3 rows of
2 stools

$6 \div 1 = 6$
1 row of
6 stools

$6 \div 6 = 1$
6 rows of
1 stool

Try These

1. Use the array to complete the sentence.

   a) $18 \div 6 = \underline{\quad 3 \quad}$

   b) $14 \div 2 = \underline{\quad 7 \quad}$

   c) $15 \div 3 = \underline{\quad 5 \quad}$

1. Write a division sentence for each array.

a)	b)	c)
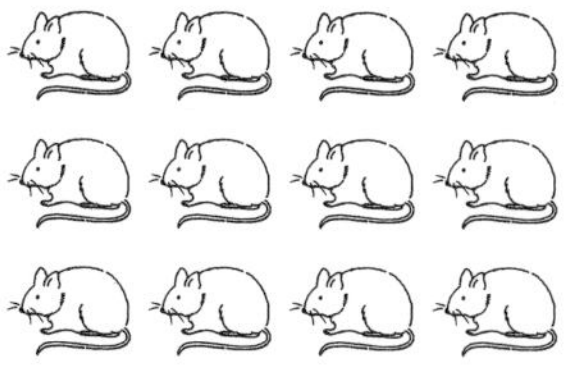 12 ÷ 3 = 4	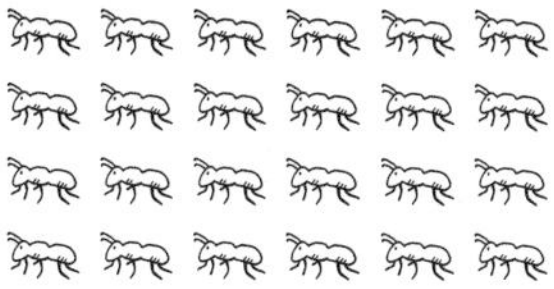 24 ÷ 4 = 6	 4 ÷ 1 = 4

2. Draw an array for each division sentence.

a) 15 ÷ 5 = **3**   b) 12 ÷ 2 = **6**   c) 24 ÷ 6 = **4**

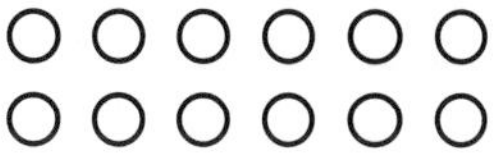

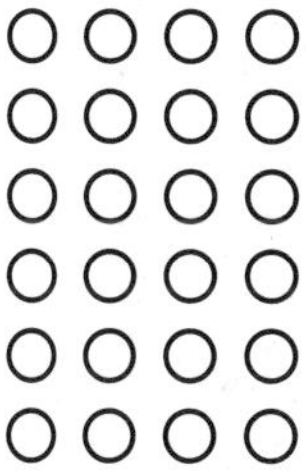

3. Use counters. Make an array to find each answer.

a) 20 ÷ 4 = **5**   b) 16 ÷ 2 = **8**   c) 6 ÷ 1 = **6**

d) 18 ÷ 9 = **2**   e) 30 ÷ 5 = **6**   f) 28 ÷ 7 = **4**

Stretch Your Thinking

There are 24 members in the Boy Scout troop.
They will march in the parade in equal rows.
How many Boy Scouts could be in each row?
Find as many answers as you can.

1, 2, 3, 4, 6, 8, 12, 24

Dividing by 2, by 5, and by 10

Quick Review

➤ Can 30 children form 5 equal groups? 2 equal groups?

30 children form
5 groups of 6.
30 is **divisible** by 5.
30 is also divisible by 6.

30 children form
2 groups of 15 children.
30 is divisible by 2.
30 is also divisible by 15.

➤ Can 14 children form groups of 5? Groups of 10?

14 is 2 groups of 5,
with 4 left over.
14 is not divisible by 5.
Groups of 5 cannot be formed.

14 is 1 group of 10,
with 4 left over.
14 is not divisible by 10.
Groups of 10 cannot be formed.

Try These

1. Use the array to help you divide.

a) $24 \div 2 = $ ___**12**___

b) $10 \div 5 = $ ___**2**___

1. Play this game with a partner.
   You will need:
   2 markers
   1 number cube labelled 1 to 6

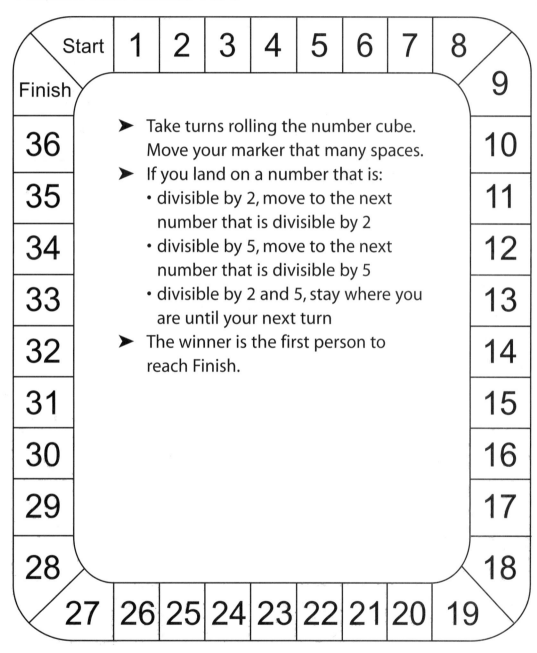

| Start | 1 | 2 | 3 | 4 | 5 | 6 | 7 | 8 |

➤ Take turns rolling the number cube. Move your marker that many spaces.
➤ If you land on a number that is:
  • divisible by 2, move to the next number that is divisible by 2
  • divisible by 5, move to the next number that is divisible by 5
  • divisible by 2 and 5, stay where you are until your next turn
➤ The winner is the first person to reach Finish.

Finish
36 35 34 33 32 31 30 29 28
27 26 25 24 23 22 21 20 19
9 10 11 12 13 14 15 16 17 18

Stretch Your Thinking

Name five numbers that are divisible by 2, by 5, and by 10.

Sample Answer: 10, 20, 30, 40, 50

Relating Multiplication and Division

Quick Review

This array has 6 rows of 7.
The multiplication sentence is: $6 \times 7 = 42$
The division sentence is: $42 \div 6 = 7$

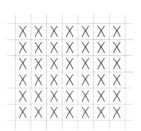

Turn the array to show 7 rows of 6.
The multiplication sentence is: $7 \times 6 = 42$
The division sentence is: $42 \div 7 = 6$

These four number sentences are **related facts**.

$6 \times 7 = 42$
$7 \times 6 = 42$
$42 \div 6 = 7$
$42 \div 7 = 6$

When you know one number fact, you can write related facts.

Try These

1. Write a multiplication sentence and a division sentence for each.

Sample Answers

a)

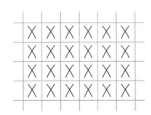

$\underline{4 \times 6 = 24 \quad 24 \div 4 = 6}$

b)

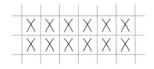

$\underline{2 \times 6 = 12 \quad 12 \div 2 = 6}$

1. Write a multiplication sentence and a division sentence for each.
   Sample Answers:

a)	b)	c)
⭐⭐⭐⭐⭐ ⭐⭐⭐⭐⭐ ⭐⭐⭐⭐⭐ ⭐⭐⭐⭐⭐ **5 × 4 = 20** **20 ÷ 5 = 4**	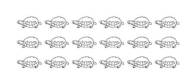 **3 × 6 = 18** **18 ÷ 3 = 6**	 **3 × 5 = 15** **15 ÷ 3 = 5**

2. Write the related facts for each set of numbers.

   a) 4, 6, 24   **4 × 6 = 24   6 × 4 = 24   24 ÷ 6 = 4   24 ÷ 4 = 6**

   b) 3, 4, 12   **3 × 4 = 12   4 × 3 = 12   12 ÷ 4 = 3   12 ÷ 3 = 4**

   c) 4, 7, 28   **4 × 7 = 28   7 × 4 = 28   28 ÷ 7 = 4   28 ÷ 4 = 7**

   d) 7, 7, 49   **7 × 7 = 49   49 ÷ 7 = 7**

   e) 7, 1, 7   **7 × 1 = 7   1 × 7 = 7   7 ÷ 1= 7   7 ÷ 7 = 1**

3. Divide. Use multiplication facts to help you.

   a) 12 ÷ 6 = **2**   b) 30 ÷ 5 = **6**   c) 21 ÷ 3 = **7**

   d) 35 ÷ 7 = **5**   e) 16 ÷ 4 = **4**   f) 7 ÷ 7 = **1**

Stretch Your Thinking

Berta has a collection of antique dolls.
If Berta puts her dolls into groups of 4 or 7, she has 2 dolls left over.
How many dolls might Berta have?

Sample Answer: 30, 58, 86

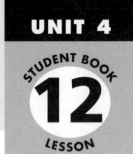
Number Patterns on a Calculator

Quick Review

At Home
At School

➤ A calculator can do repeated addition.
This is counting on.

These key strokes are for the TI-108 calculator.

Press: ON/C 4 + = = = = = = = = = =

You will see: 4 8 12 16 20 24 28 32 36 40

This is 4 multiplied by: 1 2 3 4 5 6 7 8 9 10

There is a pattern in the ones digits: 4, 8, 2, 6, 0, 4, 8, 2, 6, 0

➤ A calculator can do repeated subtraction.
This is counting back.

Press: ON/C 4 9 – 7 = = = = = = = =

You will see: 49   42 35 28 21 14 7 0

➤ A calculator can do repeated multiplication.

Press: ON/C 2 × = = = = = =

You will see: 2   4 8 16 32 64

Try These

Press the keys. Write what you see on the screen.

1. ON/C 3 + = = = = = = = = = =

3, 6, 9, 12, 15, 18, 21, 24, 27, 30

2. ON/C 48 – 6 = = = = = = = =

48, 42, 36, 30, 24, 18, 12, 6, 0

Practice

1. Press the keys. Write what you see on the screen.

a) [ON/C] [8] [+] [=] [=] [=] [=] [=] [=] [=] [=] [=] [=]

   8, 16, 24, 32, 40, 48, 56, 64, 72, 80

   What pattern do you see in the ones digits?

   8, 6, 4, 2, 0, 8, 6, 4, 2, 0

b) [ON/C] [4] [×] [=] [=] [=]

   4, 16, 64, 256

   What pattern do you see in the ones digits?

   4, 6, 4, 6

c) [ON/C] [85] [−] [5] [=] [=] [=] [=] [=] [=] [=] [=]

   85, 80, 75, 70, 65, 60, 55, 50

   What pattern do you see in the ones digits?

   5, 0, 5, 0, 5, 0, 5, 0

2. Start at 72. Keep subtracting 9 until you reach 0.

   How many 9s did you subtract?   **8**

   What division sentence does this show?   **72 ÷ 9 = 8**

3. The marching band has 108 members.
   They march in rows of 9.

   How many rows does the marching band have?   **12**

   Write a number sentence to show your answer.   **108 ÷ 9 = 12**

Stretch Your Thinking

The product of 2 numbers is 54. What might the 2 numbers be?
Find as many answers as you can.

1 and 54; 2 and 27; 3 and 18; 6 and 9

Sorting by Two Attributes

Quick Review

Here are 2 ways to sort these 7 buttons.

➤ Use a chart.
The buttons are sorted by shape and size.

Size \ Shape	Round	Square	Triangular
Big	⊙⊙ ⊙⊙	⊡	△
Small	⊙	⊡	△

➤ Use a Venn diagram.
Includes Sample Answer to Question 1 below

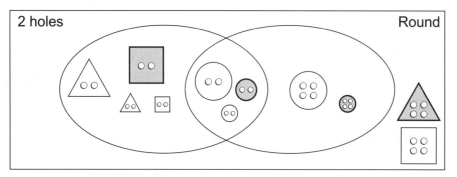

Try These

1. Draw 4 more buttons in the Venn diagram above.   **See above.**
   a) One that fits in the loop on the left
   b) One that fits in the loop on the right
   c) One that fits in the overlap
   d) One that fits outside the loops

1. a) Colour these blocks as shown.

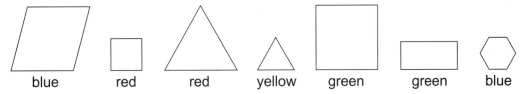

blue   red   red   yellow   green   green   blue

b) Sort the blocks above in this chart.

Size \ Colour	Red	Green	Yellow	Blue
Big	△ R	☐ G		▱ B
Small	☐ R	☐ G	△ Y	⬡ B

2. Use 2 different attributes. Sort the blocks in question 1 in this Venn diagram.

Sample Answer

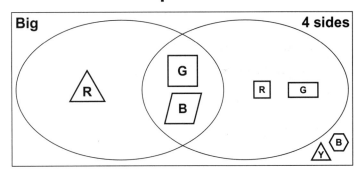

Use the Venn diagram to sort these words.

blue   jug   camp

jump   nut   jewel

old   juggle   joke

Sample Answer

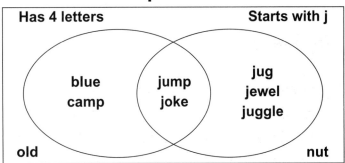

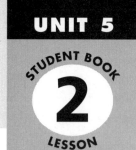

Sorting by Three Attributes

Quick Review

Look at these blocks.

Here is one way to sort them.

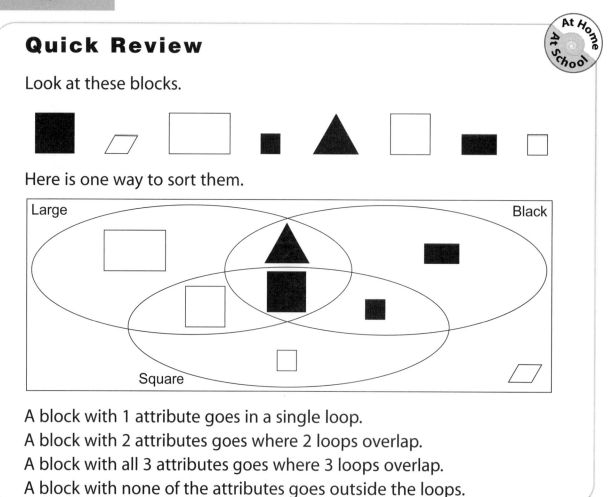

A block with 1 attribute goes in a single loop.

A block with 2 attributes goes where 2 loops overlap.

A block with all 3 attributes goes where 3 loops overlap.

A block with none of the attributes goes outside the loops.

Try These

1. Draw one block in each space in the Venn diagram.
   Sample Answer

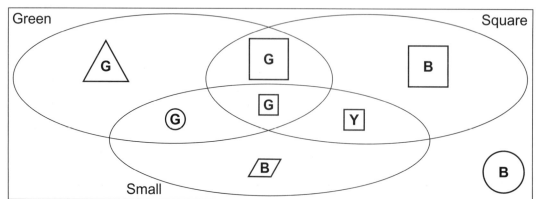

Practice

1. Colour these buttons as shown.

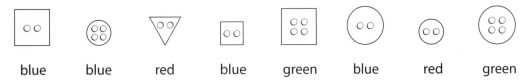

blue   blue   red   blue   green   blue   red   green

2. Sort the buttons in this Venn diagram.
   Use these attributes: Large, Round, and Blue

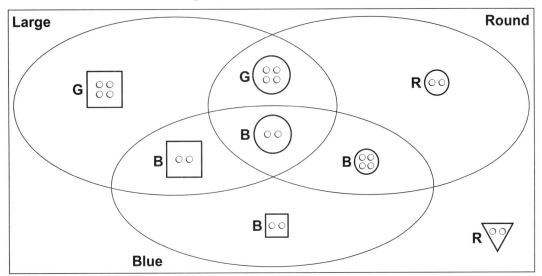

Stretch Your Thinking

Sort the buttons above using 3 different attributes. Label the loops.

Sample Answer

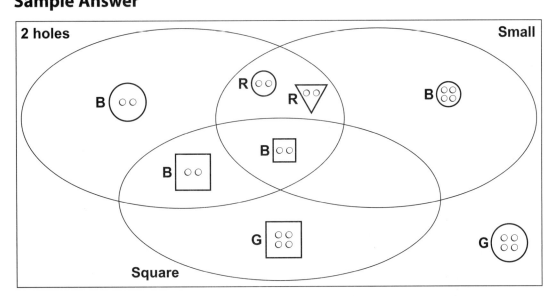

Interpreting Graphs

Quick Review

These 2 graphs show the same data.

Trees Planted in the Park

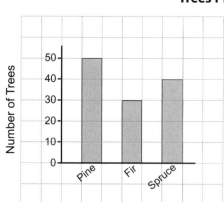

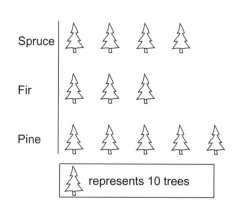

	represents 10 trees

Bar Graph   **Pictograph**

➤ You can use the bar graph to find how many pine trees were planted in the park:

The top of the pine bar lines up with 50. So, 50 pine trees were planted in the park.

➤ You can use the pictograph to find how many fir trees were planted in the park: Since 🌲 represents 10 trees, count by 10s.

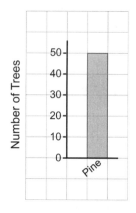

Fir   🌲   🌲   🌲

Count: 10, 20, 30

So, 30 fir trees were planted in the park.

Try These

1. Look at the graphs above.

    a) How many more pine trees than fir trees were planted?   __**20**__

    b) How many trees were planted altogether?   __**120**__

Some children were asked to name their favourite fruit.
Other children were asked to name their favourite vegetable.

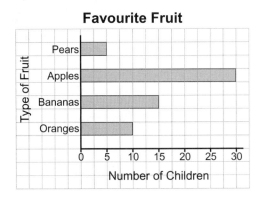

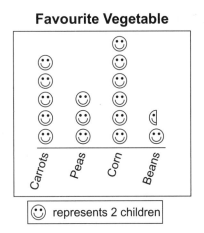

☺ represents 2 children

1. How many children like bananas? ___**15**___

2. How many more children like apples than oranges? __**20**__

3. List the fruits from least popular to most popular.

   pears, oranges, bananas, apples

4. How many children like beans? ___**3**___

5. List the vegetables from most popular to least popular.

   corn, carrots, peas, beans

6. How many children were asked to name

   a) their favourite fruit? __**60**__   b) their favourite vegetable? ___**31**___

Stretch Your Thinking

1. Use the pictograph above to answer these questions.

   a) Which vegetable was named by twice as many children as beans? **peas**

   b) Which vegetable was named by half as many children as corn? **peas**

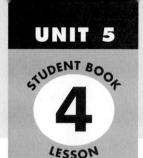

Interpreting Circle Graphs

Quick Review

These 2 graphs show the same data.

Things We Collect

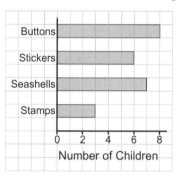

Bar Graph

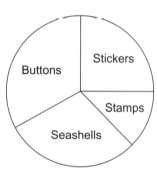

Circle Graph

➤ You can use the bar graph to find how many children collect seashells.
The end of the seashells bar is halfway between 6 and 8.
So, 7 children collect seashells.

➤ You cannot use the circle graph to find how many children
collect an item.
The circle graph shows the fraction of all the children who collect
each item. The fraction of the children who collect buttons
is about one-third.

Try These

Look at the graphs above.

1. How many children collect stickers? ____**6**____

2. About what fraction of the children collects stickers? **about one-fourth**

3. How many children collect buttons? ____**8**____

4. About what fraction of the children collect buttons? **about one-third**

1. Write 3 things you know about this graph.

   Sample Answer: About one-half of the children like summer. About one-fourth of the children like spring. About the same number of children like autumn as winter.

   Favourite Seasons

2. **a)** What does this graph show?

   favourite places to visit

   b) Where do most children like to visit?

   the zoo

   c) Which place do about one-third of the children like to visit?

   the zoo

   Places We Like to Visit

   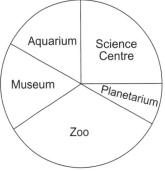

   d) Write what else you know from this graph.

   Sample Answer: About the same number of children like the aquarium as the museum. The planetarium is the least-favourite place to visit.

Stretch Your Thinking

Write 2 questions about this graph. Answer your questions.

Favourite Times of Day

Sample Answer: Which time of day do most children like? Evening

Which time of day do about one-fourth of the children like? Night or afternoon

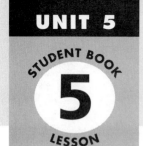
Drawing Pictographs

Quick Review

Some children chose their favourite Popsicle flavours. Here are the data.

Favourite Popsicle Flavours

Popsicle Flavours	Tally	Number of Children
Orange	✦✦✦ |||	8
Banana	✦✦✦		6
Blueberry	✦✦✦ ✦✦✦		11

To draw a pictograph:

➤ Choose a picture:

Choose how many items the picture will represent.

This is called a **key**: represents 2 children.

➤ For each flavour, draw one for every 2 children and for 1 child.

➤ Draw the pictograph.

Favourite Popsicle Flavours

Orange

Banana

Blueberry

represents 2 children

Try These

1. Draw a pictograph.
   Use 1 picture to represent 2 children.

   Favourite Kinds of Music

Kind of Music	Number of Children
Popular	9
Rock	8
Rap	10

   Sample Answer

   Favourite Kinds of Music

   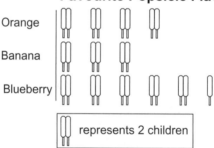

   Popular

   Rock

   Rap

   represents 2 children

104

Practice

Sample Answers

1. Meghi asked her classmates for their favourite juice. This is her tally chart.

Draw a pictograph.
Use 1 picture to represent 2 children.

Favourite Juice

Juice	Tally	Number of Children
Apple	IIII	4
Orange	IIII IIII	9
Pineapple	IIII II	7
Grape	IIII IIII	9

Favourite Juice

Apple, Orange, Pineapple, Grape

☐ represents 2 children

2. Dakota found how many children in Grades 1 to 3 signed up for T-Ball.

Draw a pictograph.
Use 1 picture to represent 2 children.

Children Who Signed Up for T-ball

Grade	Number of Children
1	20
2	15
3	10

Children Who Signed Up for T-Ball

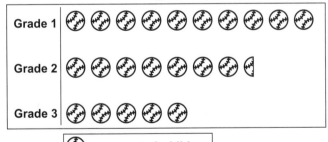

⚾ represents 2 children

Stretch Your Thinking

Suppose you used 1 picture to represent 5 children for the pictograph in question 2 above. How would that change the graph?

Sample Answer: The Grade 1 row would have 4 balls. The Grade 2 row would have 3 balls. The Grade 3 row would have 2 balls.

Drawing Bar Graphs

Quick Review

At Home At School

Draw a bar graph of the data in the chart.

➤ Choose a scale of 1 square represents 2 children. Count by 2s to find the number of squares in each bar on the graph.

➤ Write the authors on the vertical axis.

➤ Write the number of children on the horizontal axis.

Favourite Author

Author	Number of Children	Number of Squares
Judith Viorst	2	1
J. K. Rowling	12	6
Jo Ellen Bogart	8	4
Dennis Lee	10	5

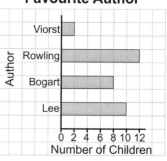

The graph shows J. K. Rowling is the favourite author. Her bar has the most squares.

Try These

The children in Mr. Shankar's class chose their favourite winter activity.

1. Finish the bar graph.

2. Write 3 things you know from the graph.

Favourite Winter Activity

Activity	Number of Children
Skating	12
Skiing	10
Sledding	8

Sample Answer: Skating is the favourite activity. Sledding is the least favourite. Four more children like skating than sledding.

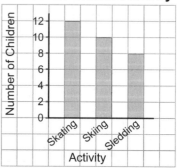

Practice

The children in four classes chose
their favourite school subject.

1. Draw a bar graph. Use a scale
   of 1 square represents 5 children.

Favourite School Subjects

Subject	Number of Children
Math	35
Science	15
Reading	25
Spelling	10
Art	15
Music	20

Sample Answer

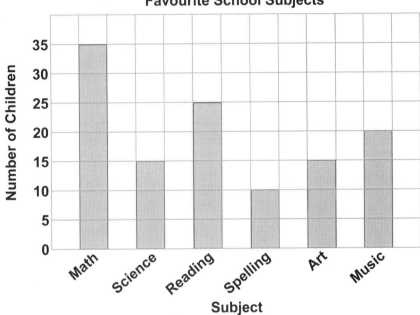

2. Write 3 things you know from the graph.

   Sample Answer: Math is the favourite subject. Spelling is the least

   favourite. Twice as many students like music as like spelling.

Stretch Your Thinking

Suppose you used a scale of 1 square represents 10 children for the graph
above. How many squares would you use for the Math bar? Explain.

I would use $3\frac{1}{2}$ squares. Three squares would represent 30 children.

One-half square would represent 5 children.

Collecting Data

Quick Review

Data are facts or information.
You collect data to learn about people and things.

➤ To collect data, begin with these questions:

- WHAT do you want to know?

- WHAT question will you ask?

- WHOM will you ask?

- HOW will you show what you find out?

You can collect data from friends and family.

➤ You can record data in tally charts and tables.

Favourite Pets	
Pet	**Number of Students**
Hamster	卌 l
Dog	卌 卌 卌 ll
Cat	卌 ll
Other	llll

Cubes We Picked Up	
Colour	**Number of Cubes**
Green	7
Red	9
Yellow	12
Blue	4

➤ You can show data in graphs.

Try These

Marsella collected these data.

1. What question do you think she asked?

   Sample Answer: Who is your

   favourite fairy tale character?

2. How many people did she ask? __33__

Character	Number of Children
Rumpelstiltskin	卌 ll
Goldilocks	lll
Cinderella	卌
Snow White	卌 卌 l
Other	卌 ll

Practice

Some children were asked how many times they have been on a train.

Times on a Train	Number of Students
0	ⲘⲘⲘⲘ II
1	IIII
2	ⲘⲘⲘ II
3	ⲘⲘⲘ ⲘⲘⲘ
More than 3	II

1. How many children have never been on a train? __12__

2. How many children were asked the question? __35__

3. Draw a bar graph to show these data.
   Sample Answer

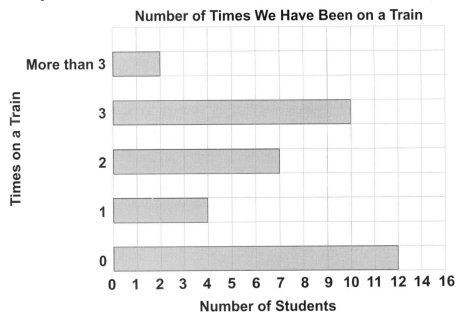

Stretch Your Thinking

Sample Answers

1. Roll a number cube 50 times. Keep a tally of the number that comes up on each roll.

2. What did you find out?

   I rolled 1 the least times.

   I rolled 5 the most times.

Number	Tally
1	ⲘⲘⲘ
2	ⲘⲘⲘ II
3	ⲘⲘⲘ II
4	ⲘⲘⲘ I
5	ⲘⲘⲘ ⲘⲘⲘ III
6	ⲘⲘⲘ ⲘⲘⲘ II

Conducting a Survey

Quick Review

At Home
At School

When you ask a question to get information, you **conduct a survey**.

Carl conducted a survey to find out what his classmates thought he should name his puppy. He gave choices with his question to help record the answers.

Carl asked:

What should I name my puppy – Harry, Sam, Barnie, or Chimo?

Each classmate answered the question.
Carl made this tally chart and bar graph:

Names for Carl's Puppy

Name	Tally	Number
Harry	|||| ||||	9
Sam	||||		6
Barnie			1
Chimo	|||| |||| ||	12

Names for Carl's Puppy

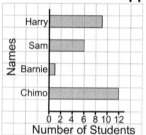

Carl found out that the most popular name was Chimo.

Try These

Sample Answers

1. Conduct a survey.
   Ask: How many pets do you have—0, 1, 2, or more than 2?
   Complete the chart.

2. Tell what you know from the chart.

Number of Pets	Tally
0				
1													
2							
More than 2	||||

The largest number of children I asked have 1 pet.

One-half of the children I asked have 1 pet.

Sample Answers

1. Conduct a survey. Ask your classmates: Would you rather fly in an airplane, a helicopter, or a hot air balloon? Complete the chart.

Favourite Ways to Fly

Vehicle	Tally	Number of Students
Airplane	ⅢⅢ ⅢⅢ ⅢⅢ ‖	17
Helicopter	ⅢⅢ	6
Hot Air Balloon	ⅢⅢ ⅢⅢ ‖	12

2. Draw a bar graph to show the data.

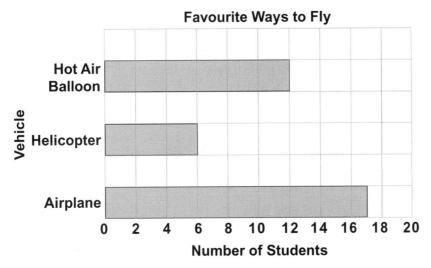

Favourite Ways to Fly

3. Describe what you found out.

   Most students would rather fly in an airplane.

   Twice as many students would rather fly in a hot air balloon

   as fly in a helicopter.

Stretch Your Thinking

1. a) Think of something you would like to find out about your classmates.
   b) Write the question you will ask.

      Sample Answer: Which sport do you like best—hockey, basketball,

      baseball, or lacrosse?

   c) Conduct the survey and record the results in a chart.
      Draw a graph to show the data.

Exploring the Calendar

Quick Review

At Home At School

JANUARY						
S	M	T	W	T	F	S
		1	2	3	4	5
1	2	3	4	5	6	7
8	9	10	11	12	13	14
15	16	17	18	19	20	21
22	23	24	25	26	27	28
29	30	31				

FEBRUARY						
S	M	T	W	T	F	S
			1	2	3	4
5	6	7	8	9	10	11
12	13	14	15	16	17	18
19	20	21	22	23	24	25
26	27	28				

MARCH						
S	M	T	W	T	F	S
			1	2	3	4
5	6	7	8	9	10	11
12	13	14	15	16	17	18
19	20	21	22	23	24	25
26	27	28	29	30	31	

APRIL						
S	M	T	W	T	F	S
						1
2	3	4	5	6	7	8
9	10	11	12	13	14	15
16	17	18	19	20	21	22
23	24	25	26	27	28	29
30						

MAY						
S	M	T	W	T	F	S
	1	2	3	4	5	6
7	8	9	10	11	12	13
14	15	16	17	18	19	20
21	22	23	24	25	26	27
28	29	30	31			

JUNE						
S	M	T	W	T	F	S
				1	2	3
4	5	6	7	8	9	10
11	12	13	14	15	16	17
18	19	20	21	22	23	24
25	26	27	28	29	30	

JULY						
S	M	T	W	T	F	S
						1
2	3	4	5	6	7	8
9	10	11	12	13	14	15
16	17	18	19	20	21	22
23	24	25	26	27	28	29
30	31					

AUGUST						
S	M	T	W	T	F	S
		1	2	3	4	5
6	7	8	9	10	11	12
13	14	15	16	17	18	19
20	21	22	23	24	25	26
27	28	29	30	31		

SEPTEMBER						
S	M	T	W	T	F	S
					1	2
3	4	5	6	7	8	9
10	11	12	13	14	15	16
17	18	19	20	21	22	23
24	25	26	27	28	29	30

OCTOBER						
S	M	T	W	T	F	S
1	2	3	4	5	6	7
8	9	10	11	12	13	14
15	16	17	18	19	20	21
22	23	24	25	26	27	28
29	30	31				

NOVEMBER						
S	M	T	W	T	F	S
			1	2	3	4
5	6	7	8	9	10	11
12	13	14	15	16	17	18
19	20	21	22	23	24	25
26	27	28	29	30		

DECEMBER						
S	M	T	W	T	F	S
					1	2
3	4	5	6	7	8	9
10	11	12	13	14	15	16
17	18	19	20	21	22	23
24	25	26	27	28	29	30
31						

Each calendar page shows the days and
weeks of one month of the year.
There are 365 days in a year.
There are 52 weeks in a year.
There are 12 months in a year.

*Every 4 years, there are
29 days in February.*

Try These

Use the calendar above.

1. **a)** How many months of the year have 30 days? _____4_____
    b) Name the months that have 30 days.

    April, June, September, November

2. Name the date that is 9 days after May 3rd. **May 12th**

Use the calendar in *Quick Review.*

1. **a)** How many Fridays are there in April?   __4__

    b) How many days are there in May and June together?   __61__

    c) Which months end on a Friday? **March and June**

    d) Which months start on a Saturday? **April and July**

    e) Which is the eleventh month? **November**

2. Name the date that is:

    a) 9 days after September 29th **October 8th**

    b) 3 weeks before July 3rd **June 12th**

    c) 6 months after March 25th **September 25th**

    d) 6 days before April 1st **March 26th**

3. A bird laid eggs on May 17th. The eggs hatched 3 weeks later.
    Name the day and the date the eggs hatched.

    Wednesday, June 7th

4. Terry's and Moe's birthdays are exactly 7 weeks apart.
    Terry's birthday is on August 19th. When might Moe's birthday be?
    Give 2 answers.

    July 1st or October 7th

Stretch Your Thinking

1. Suppose February has 29 days. Explain how you can find how many days
    there are in the year without counting or adding.

    There are normally 365 days in a year, but every 4 years there are

    29 days in February. If the calendar shows that February has 29 days,

    there are 366 days in the year.

Telling Time

At Home
At School

Quick Review

It takes 5 minutes for the minute hand to move from one number to the next number.

8 o'clock
8:00

5 minutes after 8 o'clock
8:05

This is an **analog clock**.

It is twelve fifty or ten to one.

50 minutes after 12 o'clock or
10 minutes before 1 o'clock
12:50

This is a **digital clock**.

It is five thirty-three.

33 minutes after 5 o'clock
5:33

Try These

1. Write the time on each analog clock.

a)

7:25

b)

3:40

c)

2:55

Practice

1. Write each time two ways.

a)    b)    c)

<u>**1:15**</u>   <u>**4:40**</u>   <u>**11:35**</u>

<u>**15 minutes after 1**</u>   <u>**20 minutes before 5**</u>   <u>**35 minutes after 11**</u>

2. Skip count to find how many minutes are between each pair of times.

a) 6:15 and 6:20 <u>**5 minutes**</u>   b) 8:10 and 8:40 <u>**30 minutes**</u>

c) 2:40 and 2:55 <u>**15 minutes**</u>   d) 12:00 and 12:30 <u>**30 minutes**</u>

3. Read the time on the analog clock.
Write the same time on the digital clock.

4. What is another way you could write twenty-five to seven?

<u>**6:35**</u>

Stretch Your Thinking

Lester left the library at 17 minutes before 5:00.
Draw a digital clock to show this time.

Elapsed Time

Quick Review

The amount of time from the start to the end of an activity is the **elapsed time**.

Oscar practised on his drums from 2:30 p.m. to 3:05 p.m.

To find the elapsed time in minutes, count on by 5s.

Oscar practised for 35 minutes.

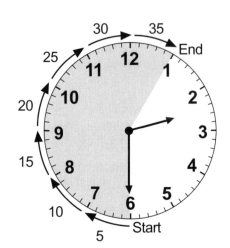

Try These

Use a clock to help you.
1. Find each elapsed time. Write the answer in minutes.

   a) 2:40 p.m. to 2:55 p.m. __**15 minutes**__

   b) 6:05 a.m. to 6:40 a.m. __**35 minutes**__

   c) 7:55 p.m. to 8:35 p.m. __**40 minutes**__

   d) 11:45 a.m. to 12:25 p.m. __**40 minutes**__

2. Tell what time it will be 25 minutes later.

   a) It's 4:30 p.m. __**4:55 p.m.**__   b) It's 1:25 p.m. __**1:50 p.m.**__

   c) It's 8:20 a.m. __**8:45 a.m.**__   d) It's 5:15 a.m. __**5:40 a.m.**__

1. Play this game with a partner.
   You will need:
   2 play clocks
   2 markers
   1 number cube

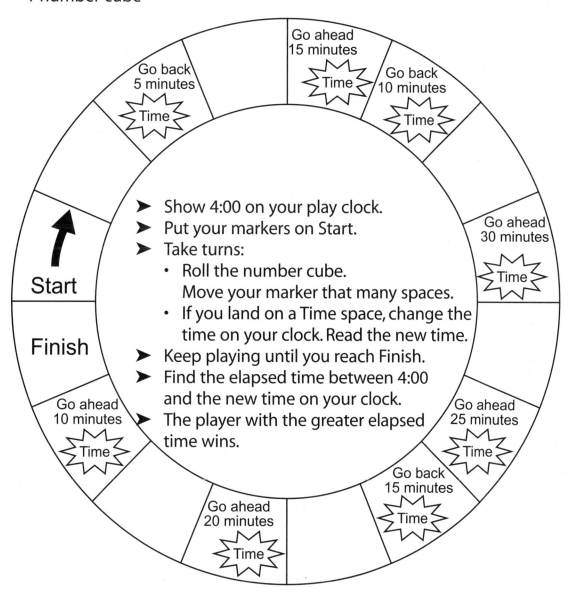

> Show 4:00 on your play clock.
> Put your markers on Start.
> Take turns:
> • Roll the number cube.
>   Move your marker that many spaces.
> • If you land on a Time space, change the time on your clock. Read the new time.
> Keep playing until you reach Finish.
> Find the elapsed time between 4:00 and the new time on your clock.
> The player with the greater elapsed time wins.

Start

Finish

Go back
5 minutes
Time

Go ahead
15 minutes
Time

Go back
10 minutes
Time

Go ahead
30 minutes
Time

Go ahead
10 minutes
Time

Go ahead
20 minutes
Time

Go ahead
25 minutes
Time

Go back
15 minutes
Time

Stretch Your Thinking

It is 11:20 p.m. What time will it be in 2 hours 25 minutes? __1:45 a.m.__

Measuring Temperature

Quick Review

You use a **thermometer** to measure **temperature**.
Temperature is measured in **degrees Celsius** (°C).

The thermometer shows a temperature of 28°C.
You say: 28 degrees Celsius

Temperatures below zero are written with a
minus sign.
You read: –5°C as "5 degrees Celsius below zero"
or as "minus 5 degrees Celsius"

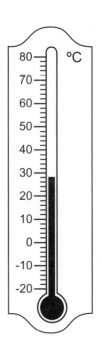

Here are some benchmark temperatures.

| Water boils | A hot day | Room temperature | Water freezes |
| 100°C | 30°C | 20°C | 0°C |

Try These

1. Circle the more likely temperature.
   a) Look at my snow fort.   (−2°C) or 19°C
   b) I don't need to wear a coat today.   3°C or (24°C)
   c) This room is warm and cozy.   (21°C) or 0°C

Practice

1. Write each temperature.

a)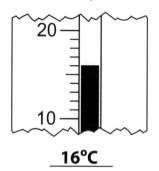

16°C

b)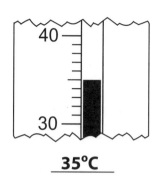

35°C

c)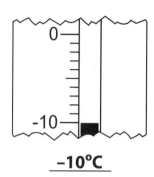

–10°C

2. Use the chart to answer each question.

a) Which was the warmest day?

Saturday

b) How much greater was the temperature on Friday than on Monday?

5°C

c) Which season do you think it is?

summer

Day	Temperature
Sunday	19°C
Monday	21°C
Tuesday	18°C
Wednesday	22°C
Thursday	27°C
Friday	26°C
Saturday	28°C

Explain. **The temperatures show that it is hot.**

3. Order the temperatures from lowest to highest.

a) –3°C, 1°C, –6°C **–6°C, –3°C, 1°C**

b) 28°C, 43°C, 36°C **28°C, 36°C, 43°C**

c) –18°C, 26°C, –14°C, 14°C **–18°C, –14°C, 14°C, 26°C**

4. Order the temperatures from highest to lowest.

a) –27°C, –2°C, –17°C **–2°C, –17°C, –27°C**

b) 0°C, 21°C, 19°C, –2°C **21°C, –19°C, –0°C, –2°C**

Stretch Your Thinking

If it is –3°C in the morning and 8°C in the afternoon, by how many degrees did the temperature increase? **11°C**

Exploring Money

At Home
At School

Quick Review

You can skip count to show how some coins and bills are related.

➤ To show how many dimes make one dollar, count on by 10s:

  10,   20,   30,   40,   50,   60,   70,   80,   90, one dollar

10 dimes make one dollar.

➤ To show how many toonies make ten dollars, count on by 2s:

   2,   4,   6,   8,   ten dollars

5 toonies make ten dollars.

Try These

1. Use play money and skip counting to help you.

   a) one loonie = ____4____ quarters   **b)** one loonie = ____10____ dimes

   c) one toonie = ____40____ nickels   **d)** one toonie = ____8____ quarters

   e) one toonie = ____20____ dimes   **f)** one loonie = ____20____ nickels

2. **a)** How many five-dollar bills would you get for a fifty-dollar bill? ____10____
   b) How many toonies would you get for a twenty-dollar bill? ____10____
   c) How many twenty-dollar bills would you get for a
   hundred-dollar bill? ____5____
   d) How many dimes would you get for a five-dollar bill? ____50____

Practice

1. How many of each coin or bill are equal to a twenty-dollar bill?
Draw pictures to show your answers.

a) toonies

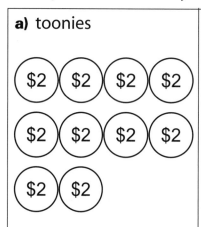

b) five-dollar bills

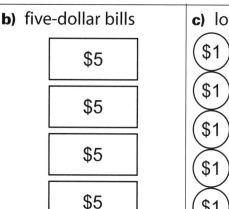

c) loonies

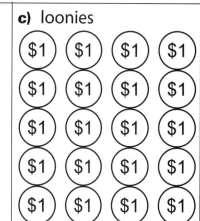

2. How many of each coin are equal to a toonie?
Draw pictures to show your answers.

a) dimes

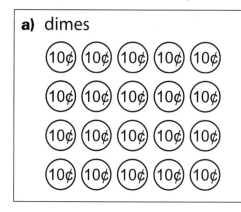

b) quarters

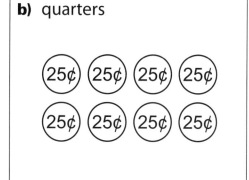

3. a) How many twenty-dollar bills make one hundred dollars? __5__

b) How many toonies would you get for 16 quarters? __2__

c) How many ten-dollar bills would you get for 25 toonies? __5__

d) How many twenty-dollar bills would you get for 20 ten-dollar bills? __10__

Stretch Your Thinking

Suppose you had 16 toonies and 18 loonies.

Which bill could you trade them for? __a fifty-dollar bill__

Estimating and Counting Money

At Home At School

Quick Review

➤ Here is one way to count a collection of money.

Sort and count the dollars.   Then sort and count the other coins.

"2,   4,   5,   6 dollars"   "25,   35,   45,   50 cents"

The total is six dollars and fifty cents, or $6.50.

dollar sign   decimal point

➤ Here are some ways you can write amounts of money.

$2.00 or $2

10¢ or $0.10

5¢ or $0.05

Try These

1. Count the money. Write each amount 2 ways.

   a)

   _____7¢ or $0.07_____

   b)

   _____$0.48 or 48¢_____

   c)

   _____$7.00 or $7_____

Practice

1. Draw pictures to show each amount.
   Sample Answers

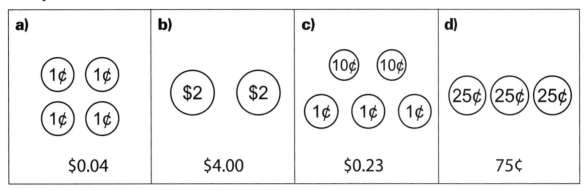

a)	b)	c)	d)
1¢ 1¢ 1¢ 1¢	$2 $2	10¢ 10¢ 1¢ 1¢ 1¢	25¢ 25¢ 25¢
$0.04	$4.00	$0.23	75¢

2. Write each amount in a different way.

   a) $0.37 __**37¢**__   b) 9¢ __**$0.09**__   c) $8.00 __**$8**__

3. Show 3 different ways to make $5.14.
   Sample Answers

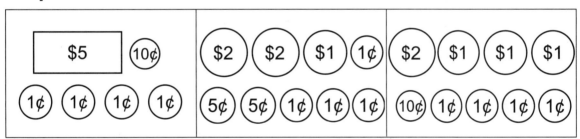

4. Tamara has 4 coins that total $0.36. Draw Tamara's coins.

Stretch Your Thinking

George has 12 coins in his pocket that total $5.60.
Draw George's coins.
Sample Answer

Making Change

Quick Review

Suppose you bought this key chain with a $5 bill.
The clerk will give you **change**.

Here is one way the clerk can make change for you.
Count on from $2.39 to $5.00:

$2.39 … $2.40, $2.50, $2.75, $3.00 $5.00

Add up the coins. Your change would be $2.61.

Try These

Draw pictures to show the money you would get for change.
Write the change in dollars.

1. You buy a bottle of water for $1.79. You pay for it with a $5 bill. **$3.21**

    (1¢) (10¢) (10¢) ($2) ($1)

2. You buy a joke book for $5.15. You pay for it with a $10 bill. **$4.85**

    (10¢) (25¢) (25¢) (25¢) ($2) ($2)

1. Draw pictures to show the money you would get for change.
   Write the change in dollars.

Item	How You Pay	Change					Amount
pen $1.19	$2 coin	1¢	5¢	25¢	25¢	25¢	$0.81
notebook $2.35	$5 bill	10¢	5¢	25¢	25¢	$2	$2.65
calculator $5.70	$10 bill	5¢	25¢	$2	$2		$4.30
markers $4.38	three $2 coins	1¢	1¢	10¢	25¢	25¢ $1	$1.62

Stretch Your Thinking

1. Tan bought a bag of hamster food for $5.09.
   He gave the clerk $10.09.

   a) Why did Tan give the clerk $10.09?

   Sample Answer: So he wouldn't get any coins back as change

   b) How much change did the clerk give Tan? **$5.00**

Adding and Subtracting Money

Quick Review

➤ Here is one way to find the total cost of the chips and drink.

Use play money to show $1.25.

Use play money to show $1.60.

Put the 2 groups of money together. Count to find the total.

The total cost of the chips and drink is $2.85.

➤ Here is one way to find how much money is left from $5.00.

Use play money to show $5.00. Take away $2.00.

Trade the loonie for other coins. Take away $0.85

There is $2.15 left.

Try These

1. Use play money to help you. How much do you have altogether?

   a) $0.60 and $1.55 __$2.15__

   b) $2.25 and $2.80 __$5.05__

   c) $5.00 and $2.13 __$7.13__

   d) $6.55 and $0.45 __$7.00__

Use play money to help you.

1. Suppose you have a $5.00 bill. Which 2 items from above could you buy?
    Sample Answers

    <u>**cards**</u>   and   <u>**dominoes**</u>

    a) Find their total cost. <u>**$3.46**</u>

    b) Find how much change you would get from $5.00. <u>**$1.54**</u>

2. Find the difference in prices between the most expensive item and the least
    expensive item.

    <u>**$5.86**</u>

3. Suppose you have a $10.00 bill.

    a) Which 3 items could you buy? <u>**Sample Answer: puzzle, checkers,**</u>

        <u>**and cards**</u>

    b) Estimate the total cost of the 3 items. <u>**$10.00**</u>

4. Suppose you buy 2 boxes of dominoes. How much change would you get
    from $5.00?

    <u>**6¢**</u>

Stretch Your Thinking

Suppose you have a $20.00 bill. Could you buy all 5 items? Explain.

<u>**Yes. I estimated $4.00 + $7.00 + $3.00 + $5.00 + $1.00 = $20.00**</u>

Exploring Capacity: The Litre

At Home At School

Quick Review

When you measure how much a container holds, you measure its **capacity**.

This bottle has a capacity of one **litre** (1 L). The bottle holds 1 L of water.

One litre fills about 4 glasses.

Here are some other things that are measured in litres.

Try These

1. Circle the containers that hold more than one litre.

  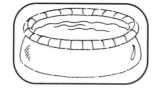

2. Circle the better estimate.

   a) (4 L) or 40 L

   b) 2 L or (20 L)
   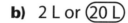

   c) (1 L) or 50 L

   d) 2 L or (200 L)

Solve each problem. Show your work.

1. How many litres of milk will it take to fill:

    a) 8 glasses? **2 L**   b) 16 glasses? **4 L**

    c) 20 glasses? **5 L**   d) 12 glasses? **3 L**

2. Eva has a 3-L jug of fruit punch.

    How many glasses can she fill? **12 glasses**

3. Each child at the picnic drank 1 glass of juice.
    There were 18 L of juice served.

    About how many children were at the picnic? **about 72 children**

4. Dakota's family drinks 4 L of milk a day.

    a) How many litres of milk does Dakota's family drink in a week? **28 L**

    b) How many litres of milk does Dakota's family drink in the month of April?

    120 L

5. Which containers hold less than 1 L? Which hold more than 1 L?

    a) a mug **less**

    b) a baby's bottle **less**

    c) a garbage can **more**

    d) a juice jug **more**

Stretch Your Thinking

1. Your heart pumps about 5 L of blood a minute.
    How many litres of blood does your heart pump in one hour?

    300 L

Exploring Capacity: The Millilitre

Quick Review

The **millilitre** (mL) is a small unit of capacity.
This teaspoon has a capacity of 5 mL.

This measuring cup has a capacity
of 500 mL. It holds 500 mL of water.

It takes 2 of these measuring cups to fill a 1-L container.

 + =

500 mL + 500 mL = 1000 mL
One litre is equal to one thousand millilitres.
1 L = 1000 mL

Try These

1. Which unit would you use to measure each capacity:
   millilitre or litre?

   a)

   b)

   c)

   millilitre   litre   millilitre

2. Peter drinks 2 L of water each day.
   How many millilitres of water does he drink each day?

   __2000 mL__

1. Circle the better estimate.

    a) (85 mL) or 850 mL   **b)** 25 mL or (250 mL)   **c)** (15 mL) or 500 mL

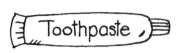   

2.

    a) Order the capacities of these containers from least to greatest.

    57 mL, 355 mL, 540 mL, 796 mL, 2 L

    b) Which container's capacity is closest to 1 L? **applesauce**

3. Which unit would you use to measure each amount: millilitre or litre?

    a) the amount of gasoline in a car **litre**

    b) the amount of water in a raindrop **millilitre**

    c) the amount of nail polish in a bottle **millilitre**

    d) the amount of water in a swimming pool **litre**

Stretch Your Thinking

It takes about 30 mL of jam to make a sandwich. About how many sandwiches could you make with this whole jar?

about 30 sandwiches

Exploring Mass: The Kilogram

Quick Review

At Home At School

When you measure how heavy an object is, you measure its **mass**.
The **kilogram** (kg) is a unit of mass.

This bag of
flour has
a mass of
about 1 kg.

This Grade 3
student has
a mass of
about 25 kg.

Try These

1. Circle the objects that have a mass of less than 1 kg.

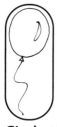     

2. Circle the better estimate.

a) (3 kg) or 60 kg   **b)** (6 kg) or 75 kg   **c)** 8 kg or (80 kg)

d) (2 kg) or 25 kg   **e)** 1 kg or (6 kg)   **f)** (1 kg) or 50 kg

132

1. Match each item with its estimated mass: 1 kg, 3 kg, 10 kg, 40 kg

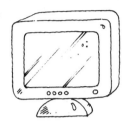 

10 kg   **40 kg**   **3 kg**   **1 kg**

2. Circle 2 objects that have about the same mass.

a)

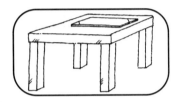  

b)

3. About how many kilograms of apples can you buy with a $10 bill?

about 5 kg

Stretch Your Thinking

Harry needs to buy 10 kg of safety salt.
Find as many ways as you can that
Harry could buy the salt.

5 kg + 5 kg   **2 kg + 2 kg + 2 kg + 2 kg + 2 kg**

3 kg + 2 kg + 5 kg   **3 kg + 3 kg + 2 kg + 2 kg**

UNIT 6

STUDENT BOOK **13** LESSON

Exploring Mass: The Gram

At Home At School

Quick Review

The **gram** (g) is a small unit of mass.
The mass of an object you can hold in the palm of your hand
is usually measured in grams.

A paper clip has a mass of
about 1 g.

Eyeglasses have a mass of
about 100 g.

It takes 1000 g to balance 1 kg.

1000 g = 1 kg

Try These

1. Circle the better estimate.

   a) 1 g or (454 g)   **b)** (5 g) or 200 g   **c)** (150 g) or 900 g

2. Match each item with its estimated mass: 1 g, 50 g, 1000 g

   a)   **b)**   **c)**

   <u>**50 g**</u>   <u>**1 g**</u>   <u>**1000 g**</u>

1. Circle the package with the greater mass.

   a)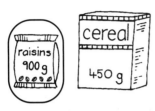

   b)

   c)

2. Which unit would you use to measure each mass:
   gram or kilogram?

   a) a penny **gram**   b) a guitar **kilogram**

   c) a pony **kilogram**   d) a box of crayons **gram**

3. Order these items from least to greatest mass.

   yogurt, cereal, margarine, nuts, sugar

4. Louanne feeds her cat 200 g of food
   each day.
   How many days will this bag of food last?

   10 days

Ross needs 2 kg of flour to make play dough.
Find 4 different ways he could buy the flour.

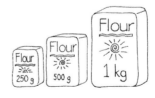

Sample Answers

1 kg + 1 kg   **250 g + 250 g + 500 g + 1 kg**

500 g + 500 g + 500 g + 500 g   **500 g + 500 g + 1 kg**

Grids and Maps

Quick Review

At Home / At School

A map is drawn on a **grid**.

To go from Min's house
to Sam's house, you move:
3 squares left and
4 squares down
or
4 squares down and
3 squares left

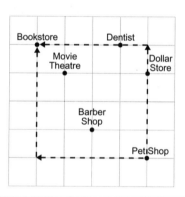

On this map, every building is at
an intersection.

To go from the pet shop to the bookstore,
you move:
4 squares left and 4 squares up
or 4 squares up and 4 squares left

Try These

1. Use the first map in *Quick Review*.
   How do you move from Sam's house to Kate's house?

   You move 2 squares up and 1 square right or 1 square right and

   2 squares up.

2. Use the second map in *Quick Review*.
   Start at the barber shop.
   Go 2 squares right, 2 squares up, and 3 squares left.

   Where do you end up? **Movie Theatre**

Sample Answer

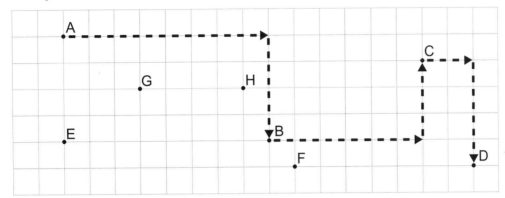

Use the grid above.

1. How do you move to go from:
   Sample Answers

   a) C to F? <u>**4 squares down and 5 squares left**</u>

   b) D to B? <u>**1 square up and 8 squares left**</u>

2. Where do you end up?

   a) You start at B, go 5 squares left and 2 squares up. <u>**G**</u>

   b) You start at A, go 7 squares right and 2 squares down. <u>**H**</u>

3. Where did you start?

   a) You moved 6 squares right and 3 squares down to get to F. <u>**G**</u>

   b) You moved 4 squares up and 8 squares left to get to A. <u>**B**</u>

Stretch Your Thinking

1. Use a coloured pencil. Draw a path on the grid in *Practice* to move from A to B, to C, to D. Do not travel on any part of the grid more than once.
   See above.

2. Describe the moves you made. <u>**Sample Answer: 8 squares right,**</u>

   <u>**4 squares down, 6 squares right, 3 squares up, 2 squares right,**</u>

   <u>**and 4 squares down**</u>

Looking at Slides

Quick Review

A **slide** moves an object along a line.
The object does not turn. The way it faces does not change.

Slides can be in different directions.

A **vertical** slide
moves a figure
up or down.

A **horizontal** slide
moves a figure
left or right.

A **diagonal** slide
moves a figure
on a slant.

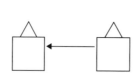

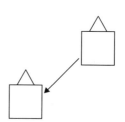

A slide is sometimes called a **translation**.

Try These

1. Circle each picture that shows a slide.

   a)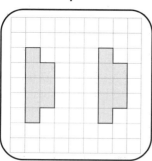

   b)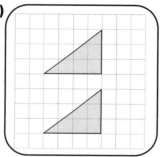

   c)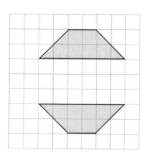

2. How do you know the pictures you circled show a slide?

   <u>**Sample Answer: The figures do not turn. The way they face**</u>

   <u>**does not change.**</u>

1. Look at the pictures below. Then answer the questions.

   A.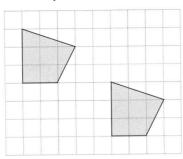

   B.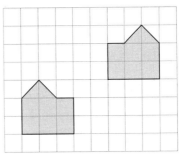

   C.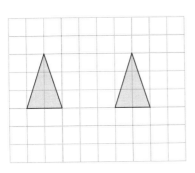

   D.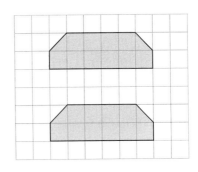

   a) Which pictures show a slide? __A, C, D__

   b) Which picture shows a vertical slide? ___D___

   c) Which picture shows a horizontal slide? ___C___

   d) Which picture shows a diagonal slide? ___A___

   e) Which picture does not show a slide? ___B___

   How do you know? __Sample Answer: The figures do not face__

   __the same way.__

Stretch Your Thinking

Put a small object on square 20.
Slide the object 4 squares left,
2 squares up, and 3 squares right.
Circle the number where the
object ended.

1	2	3	4	⑤	6	7
8	9	10	11	12	13	14
15	16	17	18	19	20	21
22	23	24	25	26	27	28

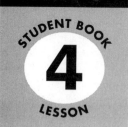
What Is a Turn?

Quick Review

From 5 o'clock to 6 o'clock, the minute hand moves **1 turn**. After 1 turn, the minute hand is back to where it started.

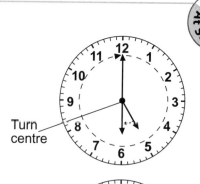

Turn centre

When the minute hand moves from 12 to 3, it moves a **quarter turn**. This is a **clockwise** turn.

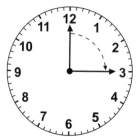

When the minute hand moves from 12 to 6, it moves a **half turn**.

When the minute hand moves from 12 to 9, it moves a **three-quarter turn**.

When the turn is the opposite way, it is a **counterclockwise turn**.

A turn is sometimes called a **rotation**.

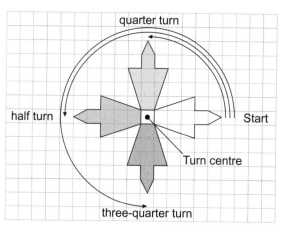

quarter turn

half turn

Start

Turn centre

three-quarter turn

Try These

1. Circle each picture that shows a turn.

a)

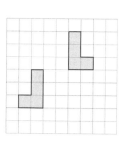

b)

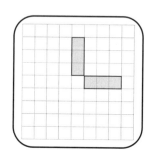

c)

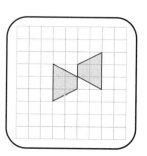

1. Look at the pictures below. Then answer the questions.

   A.

   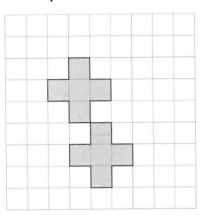

   B.

   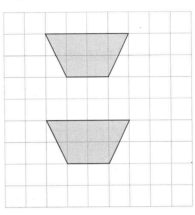

   C.

   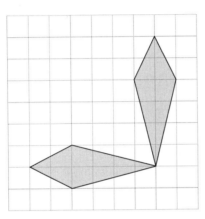

   D.

   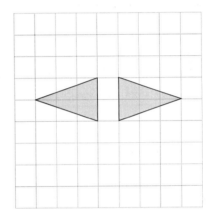

   a) Which pictures show a turn? **A, C, D**

   b) Which picture shows a quarter turn counterclockwise? **C**

   c) Which picture shows a half turn? **A and D**

   d) Which picture shows a three-quarter turn clockwise? **C**

Stretch Your Thinking

Trace a Pattern Block.
Label it "Start."
Move the Pattern Block to show a
three-quarter turn counterclockwise.
Trace the Pattern Block in its
new position.

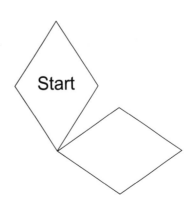

Start

Exploring Reflections

Quick Review

Each figure and its **image** show a **reflection**.

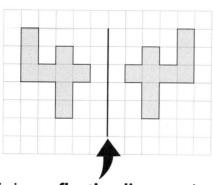
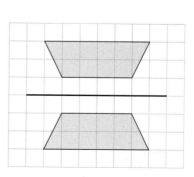

This is a **reflection line** or mirror line.

A figure and its image face opposite ways.

A reflection is sometimes called a **flip**.

Try These

1. Circle each picture that shows a reflection.
   Then draw the mirror line on the picture.

   a)

   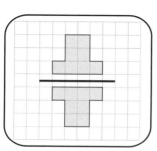

   b)

   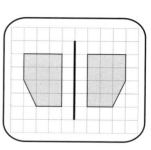

   c)

   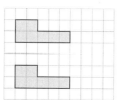

   d)

   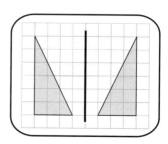

1. Circle the pictures that show a reflection.

a)  b)  c)  d)

2. Does each picture show a reflection, a slide, or a turn?

a)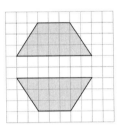

_____reflection_____

b)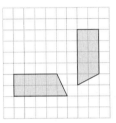

_____turn_____

c)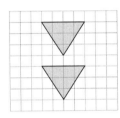

_____slide_____

d)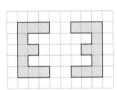

_____reflection_____

Stretch Your Thinking

1. Draw each reflection image.

a)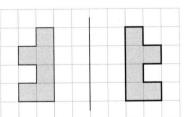

b)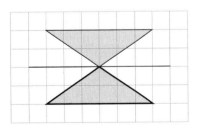

Quick Review

A **line of symmetry** divides a figure into 2 congruent parts.

You can fold along the line and the 2 parts match.

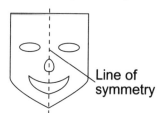

Line of symmetry

You can use a Mira to check a line of symmetry.

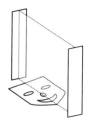

Some figures have more than 1 line of symmetry.

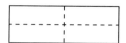

A rectangle has 2 lines of symmetry.

A regular pentagon has 5 lines of symmetry.

Try These

1. Colour the pictures that have 1 or more lines of symmetry.

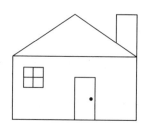

Practice

1. Label the figures below as follows:

   A – no lines of symmetry
   C – 2 lines of symmetry

   B – 1 line of symmetry
   D – more than 2 lines of symmetry

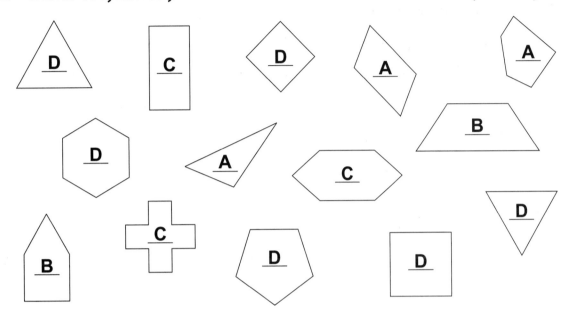

2. Look at these numbers.

   0 1 2 3 4 5 6 7 8 9

   a) Which numbers have no lines of symmetry? __2, 4, 5, 6, 7, 9__

   b) Which numbers have 1 line of symmetry? ____3____

   c) Which numbers have more than 1 line of symmetry? __0, 1, 8__

Stretch Your Thinking

1. Does a circle have more than 1 line of symmetry? Explain.

   Sample Answer: Yes. Every line drawn through the centre of a circle is

   a line of symmetry.

Exploring Equal Parts

Quick Review

Here are some ways to divide **1 whole** into **equal parts**.

4 equal parts

5 equal parts

10 equal parts

4 fourths
or **4 quarters**

5 fifths

10 tenths

Try These

1. Does each figure show equal parts? Circle *Yes* or *No*.

a)

b)

c)

(Yes)   No

Yes   (No)

(Yes)   No

2. Name the equal parts of each whole.

a)

b)

c)

fourths or quarters   **halves**   **thirds**

3. Divide each figure to show equal parts.
Sample Answers

a)

b)

c)

3 thirds

2 halves

4 fourths

146

Practice

1. Circle the figures that show equal parts.

a) 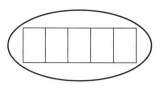 b)  c)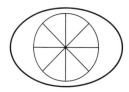

2. Name the equal parts of each whole.

a)   b) 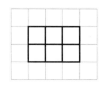 c)

   tenths   sixths   fourths or quarters

3. Divide each figure to show equal parts. Show 2 different ways.
   Sample Answers

Equal Parts	First Way	Second Way
Halves	⊖ circle with horizontal line	⊖ circle with vertical line
Quarters	square divided in 4	square divided in 4 vertical strips
Eighths	rectangle in 8	rectangle in 8 vertical strips

Stretch Your Thinking

This rectangle shows thirds.
Make it show sixths.
Sample Answer

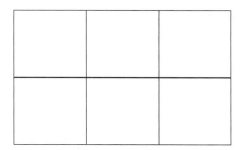

Exploring Fractions of a Length

Quick Review

You can fold a strip of paper to show fractions.

4 fourths make 1 whole.

3 thirds make 1 whole.

This strip shows tenths because all the parts are equal and there are 10 of them.

10 tenths make 1 whole.

Once you divide the length into equal parts, you can count the parts.

2 fifths are shaded.   3 fifths are not shaded.

Try These

1. What fraction of each strip is shaded?

   a)

   3 fourths or 3 quarters

   b)

   4 tenths

   c)

   6 sixths

1. Colour to show each fraction.

   Sample Answers

   a) 2 thirds

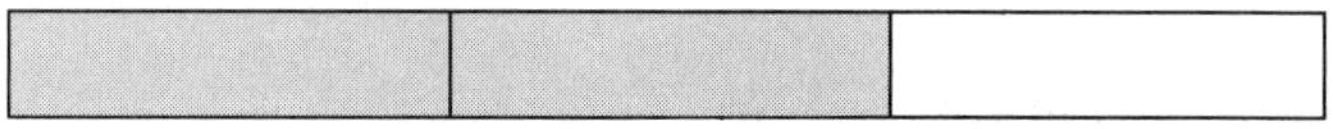

   b) 5 eighths

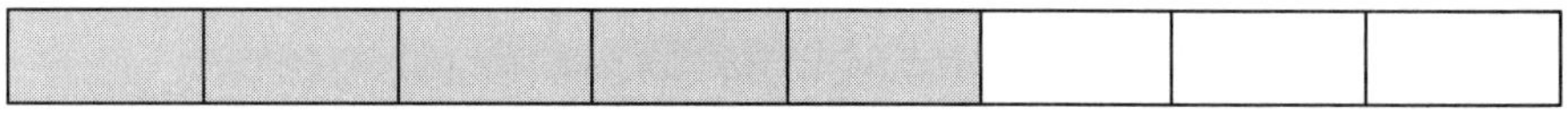

   c) 3 fifths

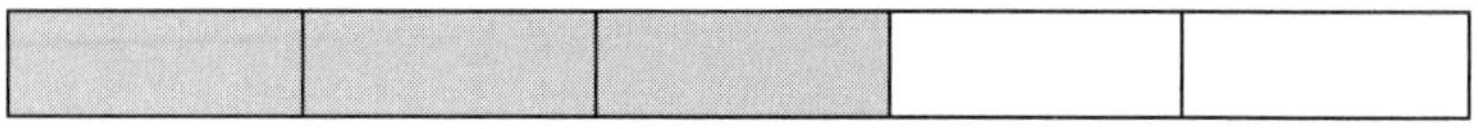

2. Estimate. About how far up the flagpole is each flag?

   a)

   _____ **1 half** _____

   b)

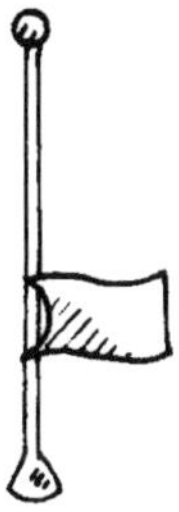

   1 third _____

   c)

   3 fourths or 3 quarters

3. Inez and Toby shared this fruit bar. Inez ate 3 eighths of the bar and Toby ate the rest. 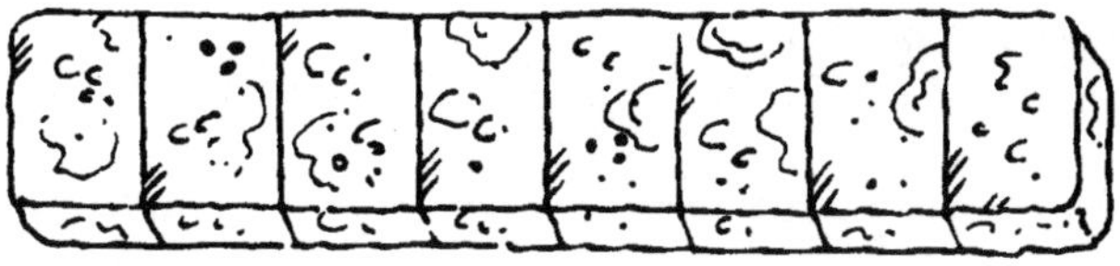
   What fraction did Toby eat? **5 eighths**

4. Estimate to colour the fraction of each strip.

   Sample Answers

   a) 1 half

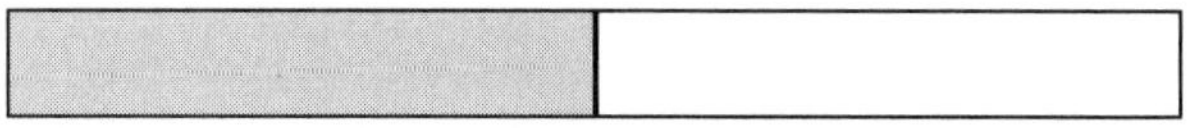

   b) 3 fourths

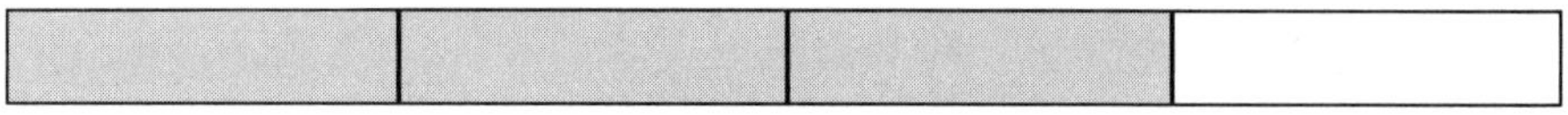

Stretch Your Thinking

Draw pictures to show how 1 quarter of a strip of paper can be longer than 3 quarters of another strip.

Sample Answer

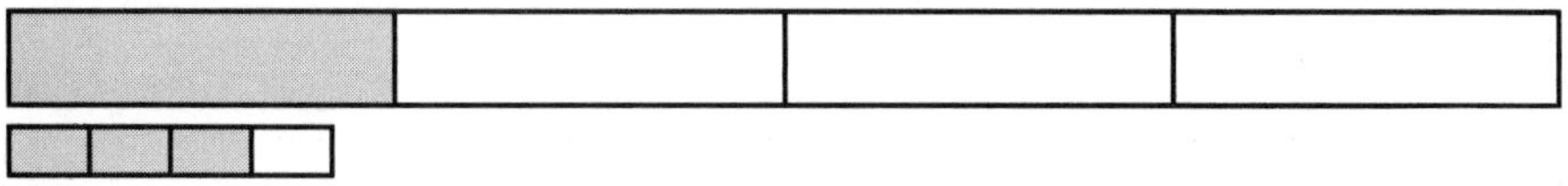

149

Exploring Fractions of a Set

Quick Review

To find a fraction of a set, start by counting.

➤ There are 8 buttons.
6 of the 8 buttons are big.
6 eighths of the buttons are big.
2 eighths of the buttons are small.

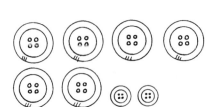

➤ There are 9 fish bowls.
7 of the 9 fish bowls have a fish.
7 ninths of the fish bowls have a fish.
2 ninths of the fish bowls are empty.

Try These

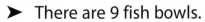

1. What fraction of each set is shaded?

   a) ○ ○ ○
      ○ ○ ○

   __4 sixths__

   b) ○ ○ ○ ○ ○ ○
      ○ ○ ○ ○ ○ ○

   __1 twelfth__

   c) ○ ○ ○ ○
      ○ ○ ○ ○

   __8 eighths__

   d) ○ ○ ○ ○

   __3 fourths or__
   __3 quarters__

2. Here are the children who
   signed up for the chess club.

   What fraction of the children are girls? __5 eighths__

   What fraction are boys? __3 eighths__

150

1. Colour some of the fish in each set.
   Write to tell what fraction you coloured.
   Sample Answers

   a)

   5 sixths

   b)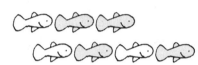

   1 tenth

   c)

   5 twelfths

   d)

   4 sevenths

2. a) Marvin has 8 pets.
      2 eighths of the pets are cats.
      3 eighths of the pets are dogs.
      The rest are hamsters.
      Draw Marvin's pets.

   b) Suppose Marvin gets 1 more cat.

      What fraction of his pets will be cats?

   3 ninths

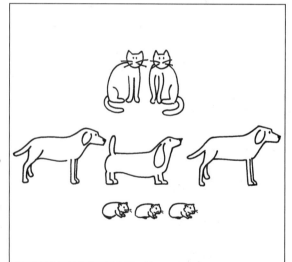

Stretch Your Thinking

Three of Sally's pencils are broken.
That's 1 quarter of Sally's pencils.
How many pencils does Sally have?
Use pictures, words, and numbers
to show your answer.

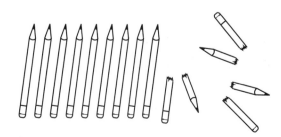

Sally has 12 pencils.

Finding a Fraction of a Set

At Home
At School

Quick Review

➤ Here are some ways to make equal groups with 10 counters.

2 equal groups of 5   5 equal groups of 2
Each group is 1 half of 10.   Each group is 1 fifth of 10.

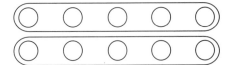

1 half of 10 = 5

1 fifth of 10 = 2

➤ To find 1 sixth of 12 counters:
Make 6 equal groups.
Each group has 2 counters.

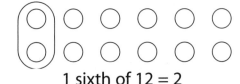

1 sixth of 12 = 2

Try These

1. Tell what fraction of each set is circled.

a)

__1 third__

b)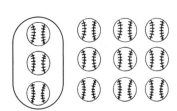

__1 fourth__

2. Draw a picture to show each fraction.
   Sample Answers

1 sixth of 6 stars   1 quarter of 8 balls

152

Practice

1. Colour to show the fraction of each set.
   Sample Answers

   a)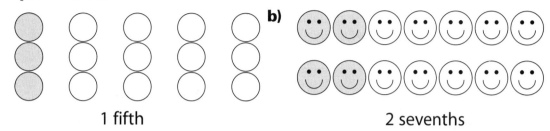

   1 fifth

   b)

   2 sevenths

2. Draw a picture to show each answer.

   a) Carmella had 12 gerbils. She gave 1 third of them to Martin.

   How many gerbils did she give to Martin? **4 gerbils**

   b) Harry drew 20 stars. He coloured 4 tenths of them yellow.

   How many stars did Harry colour yellow? **8 stars**

   c) Rapi picked 15 apples.
   1 third of the apples were red.
   The rest were green.
   How many red apples
   and how many green apples
   did Rapi pick?

   5 red apples and 10 green apples

Stretch Your Thinking

Bobby ate 1 sixth of 18 grapes. Myrna ate 1 fourth of 8 grapes.
Who ate more? Explain.

1 sixth of 18 is 3. 1 fourth of 8 is 2. Bobby ate more grapes.

Naming and Writing Fractions

Quick Review

This flag is divided into 3 equal parts, so it shows thirds.

Two of the 3 sections of the flag have stars, so the fraction is $\frac{2}{3}$.

$\dfrac{2}{3}$ ← The **top number** of a fraction tells **how many** equal parts are counted.

← The **bottom number** of a fraction tells how many equal parts are in 1 whole.

Try These ·

1. Write a fraction for each shaded part.

   a)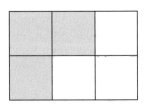

   $\dfrac{3}{6}$

   b)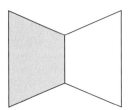

   $\dfrac{1}{2}$

   c)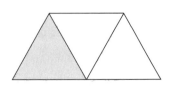

   $\dfrac{1}{3}$

2. Colour each figure to show the fraction.
   Sample Answers

   a)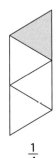

   $\dfrac{1}{4}$

   b)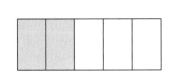

   $\dfrac{2}{5}$

   c)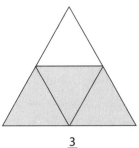

   $\dfrac{3}{4}$

1. Write a fraction for each shaded part.

a)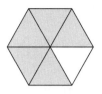

$\dfrac{5}{6}$

b)

$\dfrac{8}{8}$

c)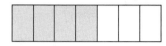

$\dfrac{4}{7}$

d)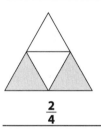

$\dfrac{2}{4}$

e)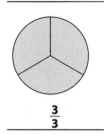

$\dfrac{3}{3}$

f)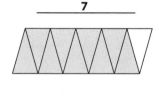

$\dfrac{9}{10}$

2. Colour each figure to show the fraction.
   Sample Answers

a)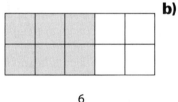

$\dfrac{6}{10}$

b)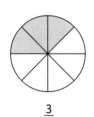

$\dfrac{3}{8}$

c)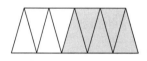

$\dfrac{5}{9}$

3. Colour the sections of this quilt.
   Use 4 different colours.
   Use fractions to describe the quilt.

   Sample Answer: $\dfrac{2}{9}$ **red,** $\dfrac{2}{9}$ **green,**

   $\dfrac{2}{9}$ **blue,** $\dfrac{3}{9}$ **yellow**

R	R	B
G	G	B
Y	Y	Y

Stretch Your Thinking

This figure represents $\dfrac{1}{3}$ of a whole.
Show what the whole might look like.
Sample Answer

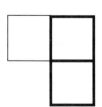

Mixed Numbers

Quick Review

You can use whole numbers and fractions to name amounts greater than 1. Here are 9 half-sandwiches.

This is $\frac{9}{2}$ sandwiches, or 4 and $\frac{1}{2}$ sandwiches.
You write 4 and $\frac{1}{2}$ as **4$\frac{1}{2}$**.
4$\frac{1}{2}$ is a **mixed number**.

Try These

1. Think of the yellow Pattern Block as 1 whole.
   Write a fraction and a mixed number
   to describe each set of Pattern Blocks.

   a)

   $\frac{3}{2}$ _____   1$\frac{1}{2}$ _____

   b)

   $\frac{9}{6}$ _____   1$\frac{3}{6}$ _____

   c)

   $\frac{7}{3}$ _____   2$\frac{1}{3}$ _____

2. Draw a picture to show each mixed number.
   a) 1$\frac{1}{4}$ b) 3$\frac{1}{2}$ c) 2$\frac{1}{3}$
   Sample Answers

   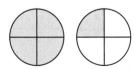   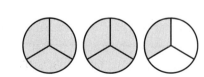

1. Write a fraction and a mixed number for each picture.

	Fraction	Mixed Number
	$\frac{13}{2}$	$6\frac{1}{2}$
	$\frac{23}{4}$	$5\frac{3}{4}$
	$\frac{11}{8}$	$1\frac{3}{8}$

2. Use pictures, numbers, and words to answer each question.

   a) Each pizza was cut into 6 equal slices.
   After the party, $2\frac{5}{6}$ pizzas were left.
   How many slices is that?

   <u>**17 slices**</u>

   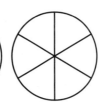

   b) Draw a picture to show $3\frac{2}{5}$.
   Sample Answer

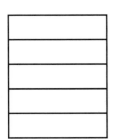

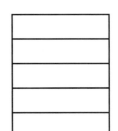
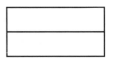

Write a fraction and a mixed number for each.

How many weeks are in 10 days? $\frac{10}{7}$ $1\frac{3}{7}$

How many years are in 17 months? $\frac{17}{12}$ $1\frac{5}{12}$

Measuring Length in Centimetres

Quick Review

A **centimetre** (cm) is a unit of **length**.

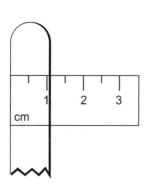

A Popsicle stick is about
1 cm wide.

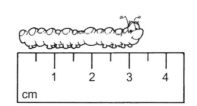

This caterpillar is between
3 cm and 4 cm long.
The length is closer to 3 cm
than to 4 cm.
The caterpillar is about 3 cm long,
or 3 cm to the nearest centimetre.

Try These

1. Draw a line to match each item with its estimate.
   a) the length of a ladybug   about 10 cm
   b) the width of your math book   about 100 cm
   c) the length of a Popsicle stick   about 1 cm
   d) the width of the classroom window   about 22 cm

2. a) Estimate the length of this line.

   _____

   Estimate: **Sample Answer: 8 cm**
   b) Measure the line to the nearest centimetre.
      Length: **10 cm**

1. Play this game with a partner.
    You will need:
    2 markers
    a collection of objects (e.g., a straw, a stapler, a shoe, a pencil)

Place your markers on Start.

➤ Player 1: Choose an object and estimate its length.

➤ Player 2: Measure the object to the nearest centimetre. Find the difference between your opponent's estimate and your measurement. Move your marker that many spaces on the game board.

➤ Switch roles and continue playing.
The first player to reach Finish is the winner.

Play the game again. Use a different collection of objects.

Start Finish

Stretch Your Thinking

Estimate the length of a line of 25 paper clips.
Then measure a line of 25 paper clips. Record your results.
Sample Answers

Estimate: __75 cm__   Measurement: __80 cm__

Measuring Length in Metres

Quick Review

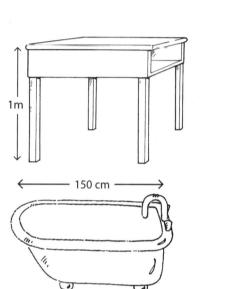

One **metre** (m) is a length of 100 cm.

1 m = 100 cm

A desk is about 1 m tall.

1m

A bathtub is about 150 cm long.
You can write this as 1 m 50 cm.

150 cm

At Home
At School

Try These

1. Draw a line to match each item with its estimate.

   a) the height of a classroom wall — about 20 m

   b) the width of a hockey net — about 3 m

   c) the length of a school hallway — about 5 m

   d) the height of a giraffe — about 1 m

2. Would you measure each item in centimetres or metres?

   a) the height of a flagpole — **metres**

   b) the length of a mouse — **centimetres**

   c) the length of a whale — **metres**

   d) the width of a hand — **centimetres**

Practice

1. Find an object that fits each description.
   Measure each object.
   Record the measurement in metres and centimetres.
   Complete the chart.

Description	Object	Measurement
about 1m long	table	1 m 20 cm
between 1m and 2m long	bulletin board	1 m 80 cm
longer than your arm	teacher's arm	60 cm
about as tall as you	painting easel	1 m 22 cm

2. Work with a partner.

   ➤ Estimate the length 4 m.
     Put 2 pieces of tape on the floor
      about 4 m apart.
   ➤ Measure to check your estimate.
   ➤ Record the measurement
     in the chart.
   ➤ Repeat with the other lengths
     in the chart.

Estimate	Measurement
4 m	3 m 95 cm
8 m	8 m 40 cm
5 m	5 m 26 cm
9 m	8 m 72 cm
3 m	3 m 10 cm

Stretch Your Thinking

Explain how you could find the approximate length of a hallway
without using a metre stick.

Sample Answer: A baseball bat is about 1 m long. I could measure

how many bats fit in the hallway and count each as 1 m.

UNIT 9

STUDENT BOOK

3

LESSON

The Kilometre

Quick Review

Long distances are measured in **kilometres** (km).

It takes about 15 minutes
to walk 1 km.

It takes about 5 minutes
to ride a bike 1 km.

1000 metre sticks laid end-to-end would stretch 1 km.

1000 m = 1 km

Try These

1. Is each distance longer than 1 km or
   shorter than 1 km?
   Sample Answers

   a) from your school to the next town or city **longer**

   b) from Vancouver to Calgary **longer**

   c) the distance of a 2-minute walk **shorter**

2. Would you measure each item in
   centimetres, metres, or kilometres?

   a) the distance of a 30-minute walk **kilometres**

   b) the length of a pony tail **centimetres**

   c) the height of a building **metres**

   d) the distance of a sack race **metres**

   e) the distance of a long jump **metres**

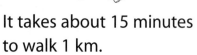

1. Draw a line to match each item with its estimate.

    a) the distance of a train ride   about 2 km

    b) the length of a blue whale   about 1 m

    c) the height of a desk   about 150 km

    d) the distance of a 10-minute bike ride   about 35 m

2. What do you think these signs mean?
   Sample Answers

   a)

   **WATERLOO
   78 km**

   <u>Waterloo is 78 km away.</u>

   b)

   Speed
   Limit
   100
   km/h

   <u>You can drive 100 km an</u>

   <u>hour on the road.</u>

3. Describe how you could measure a distance of about 1 km in your neighbourhood.

   Sample Answer: I could take a walk for 15 minutes.

   The distance I walk would be about 1 km.

Is a walk around this park less than or greater than 3 km? Explain.

Sample Answer: 3 km is 3000 m.

The distance around the park is

2600 m. That is less than

3000 m, so it is less than 3 km.

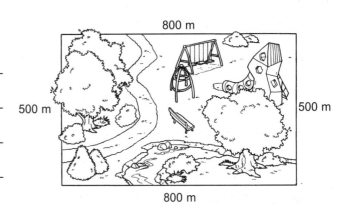

800 m

500 m   500 m

800 m

Measuring Perimeter in Centimetres

Quick Review

➤ To find the perimeter of a figure drawn on 1-cm grid paper, count the units along the outside of the figure.

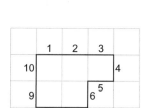

The perimeter is 10 cm.

➤ To find the perimeter of a figure not drawn on grid paper, use a ruler. Measure each side. Then add the lengths.

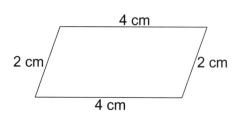

4 cm + 2 cm + 4 cm + 2 cm = 12 cm
The perimeter is 12 cm.

Try These

1. Find the perimeter of each figure on 1-cm grid paper.

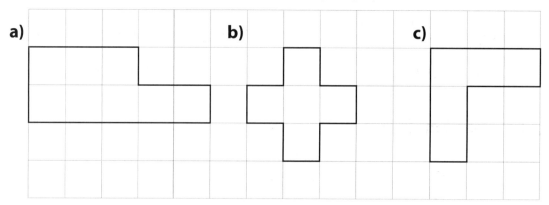

a)

b)

c)

Perimeter: **14 cm**   Perimeter: **12 cm**   Perimeter: **12 cm**

2. Measure the perimeter of this figure.

Perimeter: **16 cm**

1. Work with a partner. Take turns.
   Choose a figure. Estimate its perimeter.
   Then measure and record the perimeter.
   Sample Answers

a)

Estimate: __12__ cm

Perimeter: __12__ cm

b)

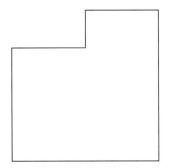

Estimate: __18__ cm

Perimeter: __16__ cm

c)

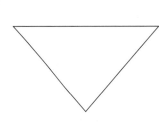

Estimate: __9__ cm

Perimeter: __10__ cm

d)

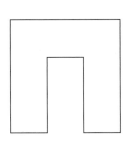

Estimate: __19__ cm

Perimeter: __16__ cm

e)

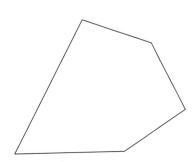

Estimate: __14__ cm

Perimeter: __13__ cm

f)

Estimate: __12__ cm

Perimeter: __12__ cm

Stretch Your Thinking

These 2 figures have different perimeters. Change one of them so that their perimeters are equal.

Sample Answer

Measuring Perimeter in Metres

Quick Review

You can use metres to measure the perimeter of a large figure.

10 m

5 m   5 m

10 m

Perimeter = 10 m + 5 m + 10 m + 5 m = 30 m
The perimeter of the garden is 30 m.

Try These

1. Find the perimeter of each figure.

a)
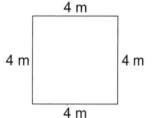
4 m
4 m   4 m
4 m

Perimeter: **16 m**

b)
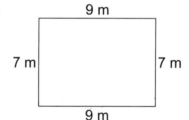
9 m
7 m   7 m
9 m

Perimeter: **32 m**

c)
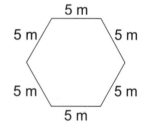
5 m
5 m   5 m
5 m   5 m
5 m

Perimeter: **30 m**

d)
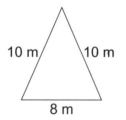
10 m   10 m
8 m

Perimeter: **28 m**

e)
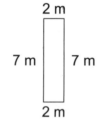
2 m
7 m   7 m
2 m

Perimeter: **18 m**

f)
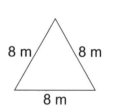
8 m   8 m
8 m

Perimeter: **24 m**

2. Use your answers from question 1.
   Order the perimeters from least to greatest.

   16 m, 18 m, 24 m, 28 m, 30 m, 32 m

1. Would you measure the perimeter of each item in centimetres or metres?

   a) a baseball diamond **metres**   **b)** a pencil case **centimetres**

   c) a page in a book **centimetres**   **d)** a farmer's field **metres**

2. Find the perimeter of each figure.
   The length of each square on the grid represents 1 m.

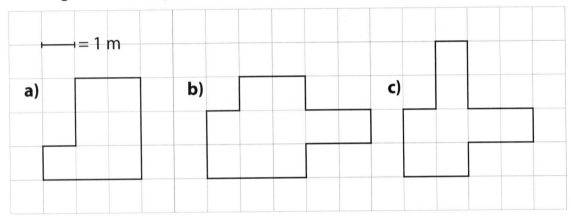

   Perimeter: **12 m**   Perimeter: **16 m**   Perimeter: **16 m**

3. The perimeter of the garden is 24 m. All the sides are equal.
   What shape might the garden be? Give as many answers as you can.
   For each answer, record the length of each side of the garden.

   Sample Answer: square, 6 m; triangle, 8 m; hexagon, 4 m;

   rhombus, 6 m

Stretch Your Thinking

Gabriel walked 100 m around the perimeter of a rectangular playground.
How long and how wide could the playground be?
Give 3 different answers.

Sample Answer: 20 m wide and 30 m long; 10 m wide and 40 m long;

15 m wide and 35 m long

Covering Figures

Quick Review

The number of units needed to cover a figure is the **area** of the figure.
The units must be the same size.
To find the area of a figure, count how many units cover it.

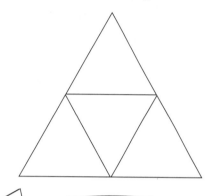

The unit is 1 green Pattern Block.
The area is 4 green Pattern Blocks.

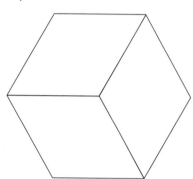

The unit is 1 blue Pattern Block.
The area is 3 blue Pattern Blocks.

Try These

1. **a)** Use yellow Pattern Blocks to
    find the area of this figure.
    Record the area in the table.

    b) Repeat using red, blue, and
    green Pattern Blocks.

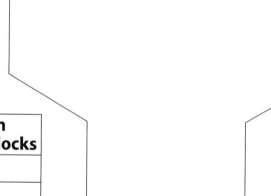

Unit	Area in Pattern Blocks
Yellow Pattern Block	3
Red Pattern Block	6
Blue Pattern Block	9
Green Pattern Block	18

1. **a)** Estimate the area of the hexagon in red Pattern Blocks.
   Then find the area in red Pattern Blocks and record it in the table.
   b) Repeat the activity with blue and green Pattern Blocks.

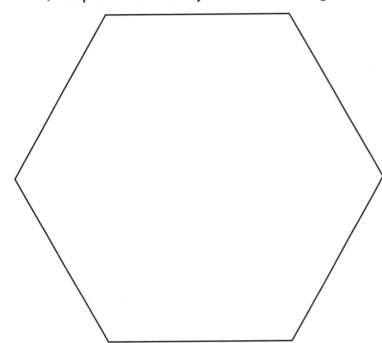

Estimates are Sample Answers

Pattern Block Unit	Estimate	Area in Pattern Blocks
red	10	8
blue	14	12
green	30	24

2. Use this grid. Draw a figure with area 3 red Pattern Blocks.

Sample Answer

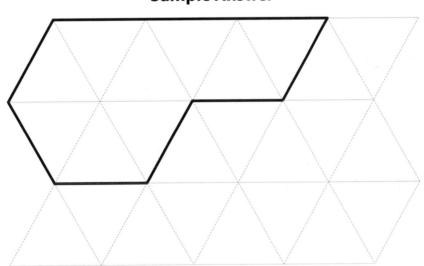

Suppose a figure has an area of 5 yellow Pattern Blocks.

What is its area in red Pattern Blocks? ___**10**___

in blue Pattern Blocks? ___**15**___

169

Measuring Area in Square Units

Quick Review

It takes 48 small squares to cover this wall.

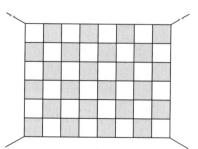

The area of this wall is 48 small square units.

It takes 12 large squares to cover this wall.

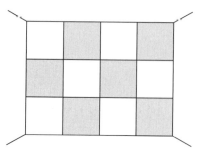

The area of this wall is 12 large square units.

When you cover a surface with small squares, the number of square units is greater than when you cover the surface with larger squares.

Try These

1. Find the area of each wall. Write your answer in square units.

a)

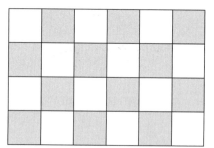

Area: __24 square units__

b)

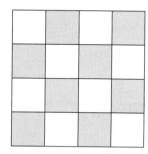

Area: __16 square units__

c)

Area: __12 square units__

d)

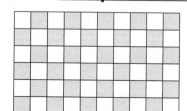

Area: __60 square units__

1. Find the area of each figure. Write your answer in square units.

a)

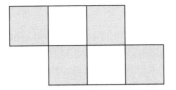

Area: __6 square units__

b)

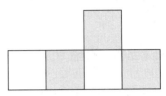

Area: __5 square units__

c)

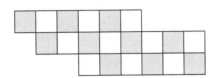

Area: __20 square units__

d)

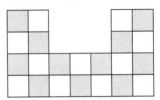

Area: __22 square units__

2. Would you use small squares or large squares to find
the area of each surface?

a) the lid of a lunch box

__small squares__

b) the gym floor

__large squares__

c) a garden

__large squares__

d) a photograph

__small squares__

Stretch Your Thinking . . . .

Suppose you covered your desktop with small squares.
Then you covered your desktop with large squares.
Would you use more small squares or more large squares? Explain.

Sample Answer: I would use more small squares because each

small square covers less space.

Using Grids to Find Area

Quick Review

Here is one way to find the area of a figure on a grid.

➤ Count the whole squares. Put an X on each square to keep track of the count.

➤ Next, count the half squares. Put a dot on each half square to keep track of the count.

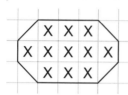

11 whole squares

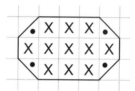

4 half squares = 2 whole squares

➤ Find the total number of squares: $11 + 2 = 13$
The area of the figure is 13 square units.

Try These

1. Find each area. Write your answers in square units.

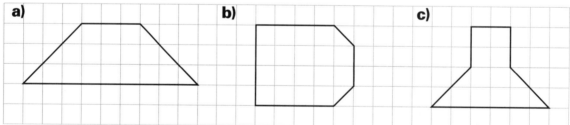

Area: __18 square units__   Area: __19 square units__   Area: __12 square units__

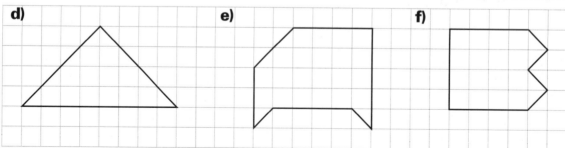

Area: __16 square units__   Area: __23 square units__   Area: __18 square units__

Sample Answers

1. Find the area of this figure.
   Then draw a different figure with the same area.

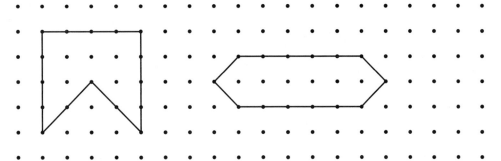

   Area: __**12 square units**__

2. Draw all the rectangles that have an area of 12 square units.

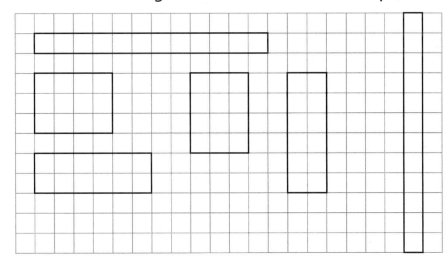

Stretch Your Thinking

1. Explain what happens to the area of a square if you double the length of its sides. Draw 2 squares to support your answer.

   Sample Answer: The area is

   4 times larger.

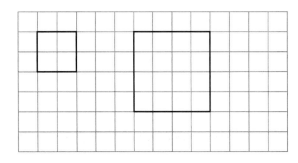

Comparing Area and Perimeter

Quick Review

➤ Different figures may have the same area.

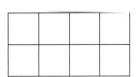

Area: 8 square units
Perimeter: 12 units

Area: 8 square units
Perimeter: 16 units

➤ Different figures may have the same perimeter.

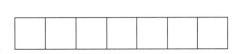

Perimeter: 16 units
Area: 7 square units

Perimeter: 16 units
Area: 16 square units

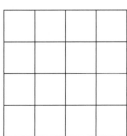

Try These

1. Find the perimeter (in units) and the area (in square units) of each figure.

a)

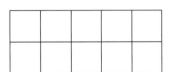

Perimeter: **14 units**

Area: **10 square units**

b)

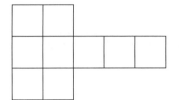

Perimeter: **16 units**

Area: **9 square units**

1. Draw 3 different figures with area 6 square units.
   Label the figures A, B, and C.
   Sample Answer

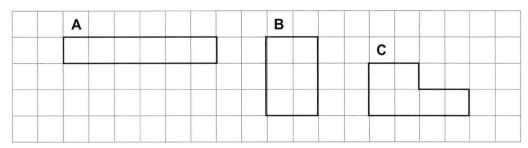

2. Use the figures you drew in question 1 to complete the chart.
   Sample Answer

Figure	Area	Perimeter
A	6 square units	14 units
B	6 square units	10 units
C	6 square units	12 units

3. Draw 2 figures with the same perimeter but different areas.
   Sample Answer

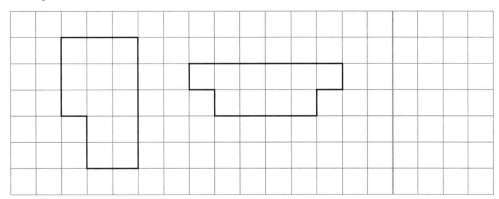

Stretch Your Thinking

Ingrid bought 12 units of fencing to enclose her yard. What might the area of her yard be? Give as many different answers as you can.

5 square units; 6 square units; 7 square units; 8 square units;

9 square units

Exploring Number Patterns

Quick Review

A **pattern rule** tells how to build a pattern.

➤ Here is a number pattern.
  2, 4, 6, 8, 10, 12, …

 This pattern rule is:

Start at 2. Add 2 each time.
The next three numbers in the pattern are 14, 16, 18.

➤ Here is another number pattern.
  47, 42, 37, 32, 27, 22, …

 This pattern rule is:

Start at 47. Subtract 5 each time.
The next three numbers in the pattern are 17, 12, 7.

➤ Here is a different number pattern.
  15, 16, 18, 21, 25, 30, …

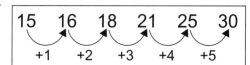

This pattern rule is:

Start at 15. Add 1. The number
you add goes up by 1 each time.

The next three numbers in this pattern are 36, 43, 51.

Try These

Fill in the missing numbers.

1. 21, 23, 25, 27, 29, __31__, __33__, __35__, __37__

2. 82, 80, 78, 76, 74, __72__, __70__, __68__, __66__

3. 50, 55, 60, 65, 70, __75__, __80__, __85__, __90__

Practice

1. Fill in the missing numbers. Write the pattern rule.

   a) 24, 25, 27, 30, 34, __39__, __45__, __52__, __60__

   Pattern rule: __Start at 24. Add 1. The number you add goes up by 1__

   __each time.__

   b) 83, 79, 75, 71, 67, __63__, __59__, __55__, __51__

   Pattern Rule: __Start at 83. Subtract 4 each time.__

2. Write the first five numbers in each pattern.
   a) Start at 14. Add 3 each time.

   __14__, __17__, __20__, __23__, __26__

   b) Start at 2. Multiply by 2 each time.

   __2__, __4__, __8__, __16__, __32__

   c) Start at 53. Subtract 5 each time.

   __53__, __48__, __43__, __38__, __33__

3. Write 2 different patterns that begin with 1, 3, … .
   Ask a friend to tell the pattern rule for each pattern.
   Sample Answers

   a) 1, 3, __5__, __7__, __9__, __11__, __13__, __15__

   b) 1, 3, __6__, __10__, __15__, __21__, __28__, __36__

Stretch Your Thinking

1. a) Complete the pattern.

   87, 86, 84, 81, 77, __72__, __66__, __59__, __51__

   b) Write the pattern rule.

   __Start at 87. Subtract 1. The number you subtract goes up by 1__

   __each time.__

Number Patterns in Tables

At Home
At School

Quick Review

Lucy walks 3 km a day.
How far does Lucy walk in 3 days? 5 days?
How many days does it take Lucy to walk 24 km?

➤ You can make a table to show the number pattern
in the kilometres Lucy walks.

The pattern rule is:

Add 3 each day.
From the table, Lucy walks
9 km in 3 days.

Days	Number of Kilometres
1	3
2	6
3	9

➤ Continue the pattern.
Keep adding 3 each day.

Days	Number of Kilometres
1	3
2	6
3	9
4	12
5	**15**
6	18
7	21
8	**24**

Lucy walks
15 km in 5 days.

It takes 8 days for
Lucy to walk 24 km.

Try These

1. Talib made this pattern with squares.

   a) Complete the table for Talib's pattern.

   b) How many squares will be in Row 6? ___**17**___ Row 8? ___**23**___

Row	Number of Squares
1	2
2	**5**
3	**8**
4	**11**
5	**14**

1. There is a pattern in each table.
    Complete each table. Write the pattern rule.

a)

Days	Pennies Saved
1	3
2	6
3	**12**
4	24
5	**48**
6	96

b)

Row	Number of Chairs
1	5
2	10
3	**15**
4	20
5	25
6	**30**

Pattern rule: **Multiply the number of pennies saved by 2 each day.**

Pattern rule: **Add 5 chairs to each row.**

2. Toya made this pattern with triangles and trapezoids. Complete the table for Toya's pattern.

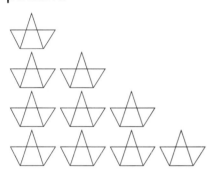

Row	Number of Triangles	Number of Trapezoids
1	**3**	**1**
2	**6**	**2**
3	**9**	**3**
4	**12**	**4**
5	**15**	**5**
6	**18**	**6**
7	**21**	**7**

Stretch Your Thinking

Here are the first 4 rows of Caleb's pattern.
How many triangles in all would Caleb need to make 6 rows? **63**

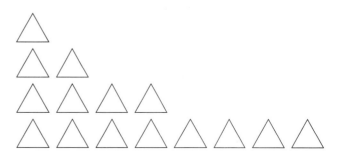

UNIT 10

STUDENT BOOK
3
LESSON

Exploring Growing Patterns

Quick Review

At Home
At School

Here is a **growing pattern**.

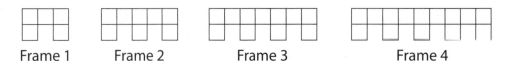

Frame 1   Frame 2   Frame 3   Frame 4

Count the number of squares in each frame.
You can show this growing pattern
in a table.

The table shows the pattern in
the number of squares in a frame.
5, 8, 11, 14, …

 The pattern rule is:
Start at 5. Add 3 each time.

Frame	Squares in the Frame
1	5
2	8
3	11
4	14

Try These

1. **a)** Use Pattern Blocks.
   Make the next three frames
   in this growing pattern.

   b) Complete the table to
   record the growing
   pattern.

   c) Write the pattern rule.

   Start at 1. Add 2 each time.

   _____

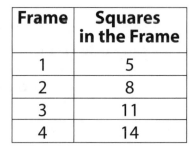

Frame 1   Frame 2   Frame 3

Frame	Triangles in the Frame
1	1
2	3
3	5
4	7
5	9
6	11

180

Practice

1. a) Use toothpicks.
Make the next 3 frames
in this growing pattern.

Frame 1　　　Frame 2　　　　Frame 3

b) Complete the table to record
the growing pattern.

Frame	Toothpicks in the Frame
1	6
2	10
3	14
4	18
5	22
6	26

c) Write the pattern rule. **Start at 6. Add 4 each time.**

2. a) Complete the table to show this growing pattern.

Frame	Squares in the Frame	Triangles in the Frame
1	1	4
2	2	6
3	3	8

Frame 1　　　Frame 2　　　Frame 3

b) How many squares would be in Frame 6?　**6**

How many triangles?　**14**

Explain how you know. **The squares increase by 1 each frame.**

The triangles increase by 2 each frame.

Stretch Your Thinking

Use toothpicks to make a growing pattern.
Draw the first 4 frames of your pattern. Write the pattern rule.
Sample Answer

   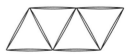

Frame 1　　　　　Frame 2　　　　　Frame 3　　　　　Frame 4

Start at 3. Add 2 each time.

Patterns with Two Attributes Changing

Quick Review

At Home
At School

Here are some **repeating patterns** in which two attributes change.

➤ In this pattern, size and position change.

The **core** of a repeating pattern is the smallest part that repeats.

The core of this pattern is:

The pattern rule is:

Large triangle, 2 small triangles turned a $\frac{1}{2}$ turn

➤ In this pattern, shape and thickness change.

The core of this pattern is:

The pattern rule is:

Thick circle, thin square

Try These

1. Which attributes change in this pattern? **size and position**

2. Draw the next 4 objects in this pattern.

1. Draw the missing objects in each pattern.

   a)

   b)

   c)

Sample Answers

2. **a)** Use buttons. Make a pattern in which two attributes change.
   Draw pictures to record the pattern.

   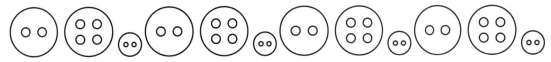

   b) Which attributes change in your button pattern?

   size and number of holes

   c) Write the pattern rule. **Large button with 2 holes,**

   large button with 4 holes, small button with 2 holes

Stretch Your Thinking

This is the core of a repeating pattern:

1. Which coin would be 11th in the pattern?   **a penny**

2. Suppose the pattern had 6 repeats of the core.
   How much money would the coins be worth?   **66 cents**

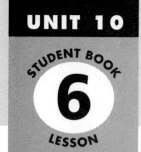

UNIT 10

Patterns with Three Attributes Changing

Quick Review

Here are some repeating patterns in which three attributes change.

➤ In this pattern, position, shape, and colour change.

The core of this pattern is:

The pattern rule is:

White triangle, two black squares, white triangle turned a $\frac{1}{2}$ turn

➤ In this pattern, colour, size, and shape change.

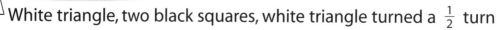

The core of this pattern is:

The pattern rule is:

Large white hexagon, small black hexagon, large white triangle, small black triangle

Try These

1. Which attributes change in this pattern? **size, shape, and position**

2. Draw the next four objects in this pattern.

184

Practice

1. Draw three cores of a repeating pattern in which:
   Sample Answers
   a) Colour, size, and shape change.

   b) Position, colour, and size change.

   c) Size, shape, and position change.

2. a) Circle the core in the following pattern.

   b) Write the pattern rule. **Large black triangle, small white triangle turned a $\frac{1}{2}$ turn, small black triangle turned a $\frac{1}{2}$ turn, large white triangle**

Stretch Your Thinking

Draw three cores of a repeating pattern in which colour, size, shape, and position change.
Sample Answer

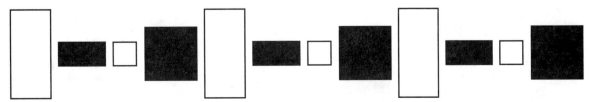

Patterns on Grids

At Home
At School

Quick Review

Here is a pattern on a grid.

➤ The symbols in the first row
are: ♣, ♦, ♥, ♣, ♦
All rows have the symbols in
the same order, but each row
starts with a different symbol.

➤ The symbols in the first
column are: ♣, ♥, ♦
All columns have the symbols
in the same order, but each
column starts with a different
symbol.

➤ To extend the pattern across,
add three more squares to
each row.
Follow the pattern: ♣, ♦, ♥

➤ To extend the pattern down,
add three more squares
to each column.
Follow the pattern: ♣, ♥, ♦

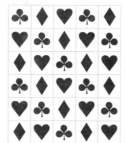

Try These

Sample Answer

1. Make a pattern on the grid. Use green, orange,
blue, purple. Write about your pattern.

   My pattern is green, green, orange, blue,

   purple, purple.

   The first and last columns are the same.

G	G	O	B	P	P	G
G	O	B	P	P	G	G
O	B	P	P	G	G	O
B	P	P	G	G	O	B
P	P	G	G	O	B	P
P	G	G	O	B	P	P

Sample Answers

1. **a)** Draw pictures to create your own pattern on the grid.

b) In which rows are the patterns the same? **1 and 6, 2 and 7, 3 and 8**

c) Write about your pattern.

The 1st, 6th, 11th, and last columns are the same.

There are four diagonals of hearts.

Stretch Your Thinking

1. **a)** Complete the pattern on the grid.

   b) Write about the patterns you see.

   Sample Answer: The 1st, 5th, and

   9th, the 2nd and 6th, the 3rd and

   7th, and the 4th and 8th columns

   are the same. The diagonals

   going down to the right have the

   pattern 7, 3, 7, 3, … .

7	3	3	7	7	3	3	7	7
3	3	7	7	3	3	7	7	3
3	7	7	3	3	7	7	3	3
7	7	3	3	7	7	3	3	7
7	3	3	7	7	3	3	7	7
3	3	7	7	3	3	7	7	3

Exploring Possible and Impossible

Quick Review

At Home At School

An **impossible** event could never happen.

You will see a walking tree.

A **possible** event could happen.

You will have homework today.

A **certain** event will definitely happen.

If you drop a ball, it will fall to the ground.

Many events are possible.

A **likely** event probably will happen.

You will read a story this week.

An **unlikely** event probably will not happen.

You will see a giraffe today.

Two **equally likely** events have the same chance of happening.

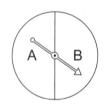

The pointer landing on A or B is equally likely.

Try These

Write whether each event is impossible, unlikely, likely, or certain.

1. You will find a loonie tomorrow. __unlikely__

2. You will see 3 suns in the sky. __impossible__

3. You will see someone you know today. __likely__

4. The sun will set in the west tonight. __certain__

Practice

1. Draw a picture to show an event for each word.
   Sample Answers

| impossible | unlikely | certain |

2. Complete each sentence.
   Sample Answers

   a) It is likely that I **will go to school tomorrow.**

   b) It is unlikely that the teacher **will ride an elephant today.**

3. Suppose you spin the pointer on this spinner.
   Complete each sentence.

   a) It is **impossible** for the pointer to land on E.

   b) It is equally likely that the pointer

   will land on **D** and **C**.

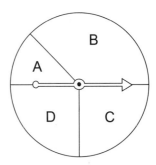

Stretch Your Thinking

Make a red, green, and yellow spinner. Color the spinner so that it is equally likely the pointer will land on red or yellow.

Sample Answer

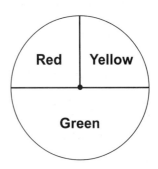

189

Conducting Experiments

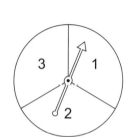

Quick Review

At Home
At School

Nori spun the pointer on this spinner
30 times.
She predicted that each number
would come up 10 times.

Nori's results are in the tally chart.
The totals for the numbers are close.

There are 3 sections on the spinner.
The sections are congruent.
So, each number has an equal chance
of coming up.

Number	Tally	Total
1	�captHℍ IIII	9
2	ℍℍ ℍℍ I	11
3	ℍℍ ℍℍ	10

Spinning a pointer is an **experiment**.
When this experiment is repeated many times, each number will likely
come up about the same number of times.

Try These

1. **a)** A penny is flipped 50 times.
   Predict how many times it will land on each side.

   Heads: ____**25**____   Tails: ____**25**____

   b) Flip a penny 50 times.
   Record the results of each flip
   in the tally chart.

   c) How do your results
   compare with your prediction?

   Sample Answer: My prediction

   and results are about the same.

   Sample Answer

Side	Tally	Total
Heads	ℍℍ ℍℍ ℍℍ ℍℍ III	23
Tails	ℍℍ ℍℍ ℍℍ ℍℍ ℍℍ II	27

Practice

1. Three cards, labelled A, B, and C, are placed in a bag.
   They are drawn one at a time and placed back in the bag after each draw.
   a) Predict how many times each letter will likely be drawn in 30 tries.

   A: __10__   B: __10__   C: __10__

   b) Conduct the experiment.
      Draw 30 cards and record
      the results on the tally chart.

   Sample Answer

Letter	Tally	Total
A	ⅢⅢ ⅢⅢ ‖	12
B	ⅢⅢ ⅢⅢ	10
C	ⅢⅢ ‖‖‖	8

   c) How do your results compare
      with your prediction?

   Sample Answer: My prediction

   and results are about the same.

   d) Repeat the experiment. This time, draw a card 60 times.
      Predict how many times each letter will likely be drawn.

   A: __20__   B: __20__   C: __20__

   Record your results in
   this tally chart.

Letter	Tally	Total
A	ⅢⅢ ⅢⅢ ⅢⅢ ‖‖‖‖	19
B	ⅢⅢ ⅢⅢ ⅢⅢ ⅢⅢ ‖	21
C	ⅢⅢ ⅢⅢ ⅢⅢ ⅢⅢ	20

   e) Compare your results with the
      results of the first experiment.

   Sample Answer: The results

   were closer to my prediction.

Stretch Your Thinking

Suppose you picked a card only 3 times.
What do you think would happen? Explain.

Sample Answer: You might get the same letter 3 times. The more times

you repeat an experiment, the more it is likely the results will match

your prediction.

UNIT 11

STUDENT BOOK

3

LESSON

Exploring Probability

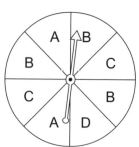

At Home At School

Quick Review

This spinner has 8 congruent parts.

The pointer is likely to land on B.
The pointer is equally likely to land on A or C.
The pointer is unlikely to land on D.
It is impossible for the pointer to land on E.

Probability tells how likely it is something will happen.

➤ 3 of 8 parts of the spinner are labelled B.
The probability of landing on B is 3 in 8.

➤ 2 of 8 parts are labelled A and 2 of 8 parts are labelled C.
The probability of landing on A is 2 in 8.
The probability of landing on C is also 2 in 8.

➤ 1 of 8 parts is labelled D.
The probability of landing on D is 1 in 8.

Try These

1. What is the probability of the pointer landing on each letter or number?

   a)

   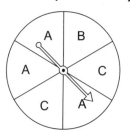

   A: __3 in 6__

   B: __1 in 6__

   C: __2 in 6__

   b)

   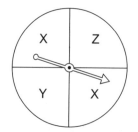

   X: __2 in 4__

   Y: __1 in 4__

   Z: __1 in 4__

   c)

   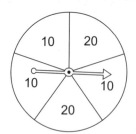

   10: __3 in 5__

   20: __2 in 5__

   30: __0 in 5__

1. a) Make a spinner with 3 colours.
        Colour it so that when the pointer
        is spun, the probability of landing
        on green is 2 in 4.
    b) What is the probability of landing
        on either of the other 2 colours?

Sample Answers

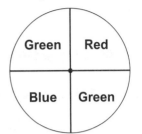

__1 in 4__

2. Draw counters in this bag.
    Colour the counters so that:
    a) The probability of picking blue is 7 in 12.
    b) The probability of picking red is 2 in 12.
    c) The probability of picking green is 3 in 12.

3. a) Colour this spinner red and blue.
        Make it so that when the pointer
        is spun, the probability of landing
        on blue is 1 in 4.
    b) What is the probability of landing on red?

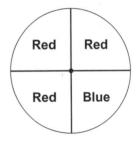

__3 in 4__

Stretch Your Thinking

Design a spinner using 3 colours.
Make it so that when the pointer is spun:

*   The probability of landing on the first
    colour is greater than the probability of
    landing on either of the other 2 colours.
*   The probability of landing on the second
    colour is greater than the probability of
    landing on the third colour.

Sample Answer

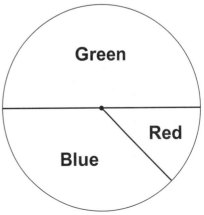

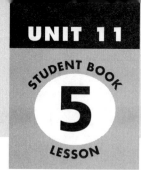

Fair and Unfair Games

At Home
At School

Quick Review

A game is **fair** if all players have an equal chance of winning.
To tell if a game is fair, find the probability of each **outcome**.

In the Spinner Game, players spin the pointer. Player A gets a point if the pointer lands on A. Player B gets a point if the pointer lands on B.

There are two possible outcomes for the spinners below.
The pointer can land on A or B.

➤ This spinner has 4 congruent parts.
The probability of landing on A is 2 in 4.
The probability of landing on B is also 2 in 4.
So, this is a fair game.

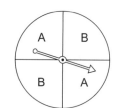

➤ This spinner has 6 congruent parts.
The probability of landing on A is 4 in 6.
The probability of landing on B is 2 in 6.
The probabilities are different.
So, this game is unfair.

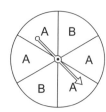

Try These

1. Tell whether each game is fair or unfair. Explain how you know.
For each game, Player A gets a point if the pointer lands on A.
Player B gets a point if the pointer lands on B.

a) 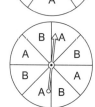   **Sample Answer: This game is unfair. The**

probability of each outcome is different.

b)   **Sample Answer: This game is fair. The**

probability of each outcome is the same.

194

Practice

1. **a)** Play this game with a partner.

   You will need: • 10 red counters   • a paper bag
   • 15 blue counters   • paper and pencils

   ➤ Put the counters in the bag.
   ➤ Take turns pulling a counter out of the bag without looking.
   ➤ Player A gets a point if the counter is red.
   Player B gets a point if the counter is blue.
   ➤ After each turn, put the counter back in the bag.
   ➤ The first player to get 10 points wins.

   b) Is this a fair game? Explain.

   No. The probability of pulling red is 10 in 25.

   The probability of pulling blue is 15 in 25.

2. **a)** Play this game with a partner.

   You will need: • a number cube   • paper and pencils

   ➤ Take turns rolling the number cube.
   ➤ Player A gets a point if an odd number shows.
   Player B gets a point if an even number shows.
   ➤ The first player to get 10 points wins.

   b) Is this a fair game? Explain.

   Yes. The probability of rolling an even or an odd number

   is the same, 3 in 6.

Stretch Your Thinking

One of the games above is unfair.
Explain what you could do to make it a fair game.

Put the same number of red and blue counters in the bag.

195

Become a Bookworm

Find a book with lots of pages (a novel would be great) and open it to the last page. Read the number.

Now, put your finger in the book where you think it's about halfway. Estimate what page number you'll see, then open it to see how close you are.

Try estimating and opening at different pages (20, 50,…)!

Story Time

Write a story that includes these two numbers:

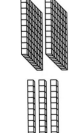

Find the Match

Which net matches the rectangular prism?

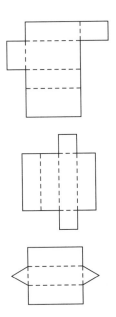

Find a similar box in your house. Cut it apart to check!

The next 4 pages fold in half to make an 8-page booklet.

Fold

Math at Home

Math is such an important part
Of everything I do.
I use it to spend my allowance
or to see how much I grew.

I use it to cook in the kitchen
Or to plan a special day.
Even racing my little brother,
Uses math in a special way!

Math at Home 1

Terrific Tens

How long would a paper chain be that is 100 links long? Let's find out!

Use 2 different colours of paper.
Cut lots of small strips 20 cm x 1.5 cm.

Glue the ends of a strip to make a circle.
Thread the next strip through the circle and glue the ends.
Continue making the chain.
Alternate colours every 10 links.

All done? Use your chain to find something in your house that is about:

26 links long   57 links long   14 links long

Game Time

Lay your paper chain along a table in a straight line.

Hide your eyes while a partner writes a word in one of the links.

Now you will need to guess where it is!

Keep guessing until you find it.
But remember … use what you already know!

Did You Know …

100 pieces of spaghetti bundled together are about as far around as 4 pencils. But, if laid end-to-end, they would be as long as 12 couches!

About how many pieces of spaghetti are the length of 1 couch?

Waiting in Line

If you were waiting in a line to get tickets to see your favourite show, what position would you rather be in?

43rd or 83rd

How many people are in line between these two spots?

When might it be best to be first? Last?

Shopping

Nancy wants to buy these 2 items.

$42.00

$23.00

She estimates how much money she will need to get from the bank machine.
She thinks, "$40 and $20 is $60, so I will need to get $60!"
Will she have enough?

What's Your Angle?

What type of angle is most popular at your house?
It's time to find out!

Set up a chart and start to search and tally!

Tally			
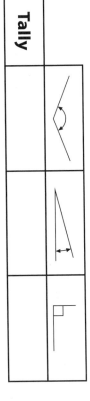

Were you surprised? Why?

Greater Than or Less Than

Take a handful of counters and put them on the table. Use craft sticks or straws to make a greater than (>) or less than (<) sign beside the pile.

A partner takes another pile of counters to make the sign true.

Together, count the piles.
Was the sign pointing the right way?

Spend It!

Before you play:

Cut out pictures from several different flyers and give each item a price (each item must be less than $25). Record prices in dollars.
Put them in a bag you can't see through.

On your turn:

At the top of a piece of paper, print $99.

Pull an item from the bag. Subtract the price from your $99. Throw the item back in the bag.

Take turns until only one person has money left.
The last player with some cash wins!

Home Design

Make a model of your home using solids.
Which solids did you use?

Draw a sketch of your model from the front, back, side, and bottom (as a worm might see it).
Which one was easiest? Why?

Secret Word

You'll need:

▲ 2 hundred charts (copy the one at the left)

▲ centimetre cubes

▲ a hard cover book for a barrier

To begin:

Players put their hundred charts in front of them. Then they place the barrier so they cannot see each other's hundred chart.

On your turn:

Spell a two-letter word on your hundred chart by covering numbers with centimetre cubes.

Now, explain to your partner where to put cubes in order to discover your secret word.

For each row of cubes, give a hint like: "Put a cube on the 12, then add 10 three times." Cover each number with a cube.

Another hint might be: "Begin at the 12, but this time add 11 three times."

Give pattern hints until the word is done. Take down the barrier to see if the secret word has been revealed!

Hundred Chart for Secret Word

1	2	3	4	5	6	7	8	9	10
11	12	13	14	15	16	17	18	19	20
21	22	23	24	25	26	27	28	29	30
31	32	33	34	35	36	37	38	39	40
41	42	43	44	45	46	47	48	49	50
51	52	53	54	55	56	57	58	59	60
61	62	63	64	65	66	67	68	69	70
71	72	73	74	75	76	77	78	79	80
81	82	83	84	85	86	87	88	89	90
91	92	93	94	95	96	97	98	99	100

Think About It!

Sam is counting coins. He says:
"25, 50, 60, 70, 80, 90, 95, 96, 97."
What coins does he have?

Mike has 38 cents in 7 coins.
What coins could they be?

At the Grocery Store

It's multiplication time! Search for items that come in groups.

6 bagels? 8 cupcakes?
4 tomatoes?

Think about how many you would have if you bought 2 packages? How about 5 or even 10 packages?

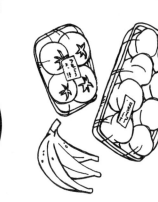

What Time Is It?

What does the clock look like at your favourite time of day?

What about your least favourite time of day? (Could it be when your alarm clock goes off?)

Fold

Math at Home

Sometimes I wonder what the world
Would be like without math.
Would we know how full to fill the tub
When it's time to take a bath?

Would we know how much paint we need
To paint my bedroom wall?
Would we know how far to throw
Fido's slimy, squishy ball?

Things would sure be different,
When it's time to make a meal.
I guess we should be thankful.
That math is very REAL!

Reflection Action

Create 2 identical designs, where each is a reflection of the other.

On the grid below, put a centimetre cube on one of the squares on the left side.

Your partner puts a cube in the square on the opposite side that is the reflection of the first.

Continue taking turns until 10 cubes have been placed on each side. What do you notice about the distance between each pair of facing cubes?

Can you make up a new game where you play until the grid is bare?

Toss and Tally

You'll need:

▲ counters
▲ a coin
▲ a number cube
▲ 6 bowls

Toss an equal number of counters into bowls and then multiply to see how many there are altogether.

On your turn:

Roll the number cube. Put that many bowls in a row, between you and your partner.

> Toss a coin. If you toss …
> heads: toss 2 counters per bowl
> tails: toss 5 counters per bowl

Together, toss the counters into the bowls. Be careful to get the right number in each one.

Whoever throws in the last counter wins that round.

Together, multiply to find how many counters are in the bowls altogether.

You decide how many rounds to play.

How Much Is a Gram?

A jellybean has a mass of about 1 g.

Search around the house for things that might be less than 1 g, about 1 g, and more than 10 g. Tally your findings.

Tally	Less than 1 g	About 1 g	More than 10 g

Now, graph what you found in 2 different ways!

Bar graph? Pictograph? Circle graph? You decide!

Mystery Number

Pick a number you think can be divided equally. Enter it into a calculator and show it to a friend.

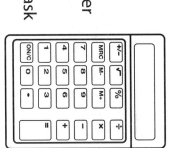

Now, secretly divide it by a number that makes equal groups.

Show the answer to a friend and ask what you did to get that number.

Traffic Jam

The car is stuck in traffic. Write directions for the driver to get out. Don't worry about which way the car is pointing, just give the directions from your point of view. You might begin by saying, "right 2 squares, down …"

Reach the arrow for the way out!

How Many in a Kilogram?

While at the grocery store, ask a family member to tell how many apples (or bananas, or …) he or she thinks would be in 1 kg. Then add apples to the scale to make that mass. How close was the estimate?

Rectangle Wrangle

The goal here is to draw the last 12-square rectangle.
Before you begin, make 5 copies of the grid (left).

On your turn:
On the grid, draw a rectangle that covers 12 squares.

Say the multiplication and division sentence
it represents.

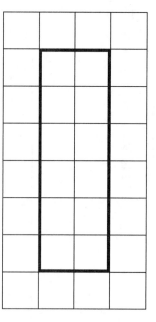

So:
$2 \times 6 = 12$
$12 \div 2 = 6$

Take turns until no more 12-square rectangles can
be made. The last person to draw a rectangle wins.

Want to play again?
Choose a different
size rectangle.

What would a 10-square
rectangle look like?
An 8-square rectangle?

Hmm … do you see a pattern?

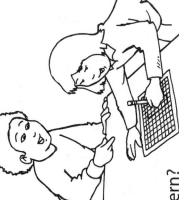

Rectangle Wrangle Grid

Fourths? Or Not?

Michel divided these 3 shapes into fourths.

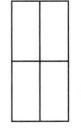

Andrea said there was a problem! What is it?

Slow Poke

At its fastest speed, a three-toed sloth travels 2 m in 1 minute. If it keeps moving forward at the same speed, how far will it go in 10 minutes?

Complete the table!

Do you see any patterns?

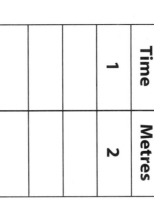

Time	Metres
1	2

Fold

Math at Home

At night I slowly close my eyes
And start to count some sheep.
You would think that very quickly,
I would fall asleep.

But I get distracted
Deciding which way is best:
By 2s, by 5s, by 10s, and more,
No wonder I get no rest!

Turbo Challenge

Be the first to get your car all the way down the hall.

You'll need:

- ▲ 2 toy cars
- ▲ 2 counters
- ▲ a coin
- ▲ a ruler
- ▲ a metre stick or metre-long string
- ▲ cards numbered 1–10 placed in a bag

On your turn:

Toss the coin.

Heads = centimetre   Tails = metre

If you toss heads, choose a number card. You'll move your car down the hall the same number of centimetres as shown on the card. If you toss tails, move your car ahead 1 m.

In either case, estimate where you will be after you move, and put a counter on that spot. Together, measure to see how close your estimate was.

Done measuring? Drive to your new spot.

Take turns until someone makes it to the end of the hall!

Tile It

Julie picked an interesting assortment of colours to tile the bathroom floor.

Follow the directions to colour the tiles.

$\frac{1}{3}$ blue $\frac{1}{9}$ green $\frac{1}{6}$ dotted

$\frac{1}{3}$ striped $\frac{1}{18}$ yellow

Now make up your own tile pattern!

Measure It

Jess said the couch was 215 cm long. Drew said it was actually 2 m 15 cm. Who's right?

It's a Deal

Make a deal with someone in your family.

"If you toss a coin and get heads, I'll do the dishes! But, if it comes up tails, you have to do them."

How likely is it that you'll have to do the dishes?

Should you make the deal?

Spot the Pattern Mix-up

Can you find the mistake in this pattern?

A	B	B	C	C
C	C	A	B	C
C	A	C	B	C
C	C	C	A	B
C	C	C	A	B
B	C	C	A	C
B	B	C	A	C
B	B	C	C	C

Is there a quick way to spot the mix-up?

Make up your own and try to stump a friend!

How Much Farther?

You can play this estimation game in the car.

Ask the driver to watch the odometer and tell you when a new kilometre is about to begin. When you think you have driven a kilometre, say "STOP!" The driver watches the odometer and lets you know when you've really travelled 1 km.

Play several times to see how close you can get.

Hmm ... How would your estimate change if you were riding a bike? Walking? Driving on the highway? In town?

All the Way Around

If one side of a square is one unit, what's the perimeter of the shape below?

_____ units

Now, use the same number of squares to make a shape that has a greater perimeter. How about one that's less?

Amazing Area Grid

Amazing Area

Find two objects smaller than the grid, and estimate which one covers the greatest area.

Put one object on the grid and trace around it. Trace the second object on top of the first in a different colour.

Count up the squares covered by the objects. Was your prediction right?

(Did you really have to count?)

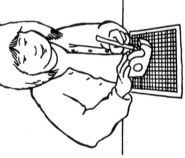

Shape Designs

Without showing anyone, draw a design using both shapes and patterns.

Now, sit back-to-back with a partner. Explain how to draw your design, while your partner draws it.

All done? Show your masterpieces! How close are they?